BRITISH RAILW

CW00742550

DMUS
RAIL SYSTEMS

FOURTEENTH EDITION
2001

The complete guide to all Diesel Multiple
Units which operate on the Railtrack network
plus the Rolling Stock of Light Rail & Metro
Systems (excluding London Underground)

Peter Fox

ISBN 1 902336 17 8

© 2000. Platform 5 Publishing Ltd., 3 Wyvern House, Sark Road, Sheffield,
S2 4HG, England.

CONTENTS

UPDATES

An update to all the books in the *British Railways Pocket Book* series is published every month in the Platform 5 magazine, *Today's Railways*, which also contains news and rolling stock information on the railways of Britain, Ireland and Continental Europe. Rolling stock updates will also be found in other magazines specialising in British matters. For further details of *Today's Railways*, please see the advertisement inside the front cover of this book.

Information in this edition is intended to illustrate the actual situation on Britain's railways, rather than necessarily agree with TOPS and computer other records. Information is updated to 13 November 2000.

READERS' COMMENTS

With such a wealth of information as contained in this book, it is inevitable a few inaccuracies may be found. The author will be pleased to receive notification from readers of any such inaccuracies, and also notification of any additional information to supplement our records and thus enhance future editions. Please send comments to: Peter Fox, Platform 5 Publishing Ltd., Wyvern House, Sark Road, Sheffield, S2 4HG, England.

Tel: 0114 255 2625 Fax: 0114 255 2471   e-mail: peter@platfive.freeserve.co.uk.

Both the author and the staff of Platform 5 regret they are unable to answer specific queries regarding locomotives and rolling stock other than through the 'Q & A' section in the Platform 5 magazine *Today's Railways*.

ORGANISATION & OPERATION OF BRITAIN'S RAILWAY SYSTEM

INFRASTRUCTURE & OPERATION

Britain's national railway infrastructure (i.e. the track, signalling, stations and associated power supply equipment) is owned by a public company – Railtrack PLC. Many stations and maintenance depots are leased to and operated by Train Operating Companies (TOCs), but some larger stations remain under Railtrack control. The only exception is the infrastructure on the Isle of Wight, which is nationally owned and is leased to the Island Line franchisee.

Trains are operated by TOCs over the Railtrack network, regulated by access agreements between the parties involved. In general, TOCs are responsible for the provision and maintenance of the locomotives, rolling stock and staff necessary for the direct operation of services, whilst Railtrack is responsible for the provision and maintenance of the infrastructure and also for staff needed to regulate the operation of services.

DOMESTIC PASSENGER TRAIN OPERATORS

The large majority of passenger trains are operated by the TOCs on fixed term franchises. Franchise expiry dates are shown in parentheses in the list of franchisees below:

Franchise	Franchisee	Trading Name
Anglia Railways	GB Railways Ltd. (until 4 April 2004)	Anglia Railways
Cardiff Railway	National Express Group PLC (until 31 March 2001)	Cardiff Railways
Central Trains	National Express Group PLC (until 1 April 2004)	Central Trains
Chiltern Railways	M40 Trains Ltd. (until 20 July 2003)	Chiltern Railways

Cross Country	Virgin Rail Group Ltd. (until 4 January 2012)	Virgin Trains
Gatwick Express	National Express Group PLC (until 27 April 2011)	Gatwick Express
Great Eastern Railway	First Group PLC (until 4 April 2004)	First Great Eastern
Great Western Trains	First Group PLC (until 3 February 2006)	First Great Western
InterCity East Coast	GNER Holdings Ltd. (until 4 April 2004)	Great North Eastern Railway
InterCity West Coast	Virgin Rail Group Ltd. (until 8 March 2012)	Virgin Trains
Island Line	Stagecoach Holdings PLC (until 12 October 2001)	Island Line
LTS Rail	National Express Group PLC (until 25 May 2011)	C2C
Merseyrail Electrics	Arriva PLC (until 17 February 2001)	Merseyrail Electrics
Midland Main Line	National Express Group PLC (until 27 April 2008)	Midland Mainline
North London Railways	National Express Group PLC (until 1 September 2004)	Silverlink Train Services
North West Regional Railways	First Group PLC (until 1 April 2004)	First North Western
Regional Railways North East	Arriva PLC (until 17 February 2001)	Northern Spirit
Scotrail	National Express Group PLC (until 30 March 2004)	ScotRail
South Central	Connex Transport UK Ltd. (until 25 May 2003)	Connex
South Eastern	Connex Transport UK Ltd. (until 12 October 2011)	Connex
South Wales & West	National Express Group PLC (until 31 March 2001)	Wales & West Passenger Trains
South West	Stagecoach Holdings PLC (until 3 February 2003)	South West Trains
Thames	Victory Railways Holdings Ltd. (until 12 April 2004)	Thames Trains
Thameslink	GOVIA Ltd. (until 1 April 2004)	Thameslink Rail
Great Northern	National Express Group PLC (until 31 May 2001)	WAGN
West Anglia	National Express Group PLC (until 4 April 2004)	WAGN

The above companies may also operate other services under 'Open Access' arrangements. The following operators run non-franchised services only:

Operator	*Trading Name*	*Route*
British Airports Authority	Heathrow Express	London Paddington–Heathrow Airport

| Hull Trains | Hull Trains | London King's Cross–Hull |
| West Coast Railway Co. | West Coast Railway | Fort William–Mallaig |

INTERNATIONAL PASSENGER OPERATIONS

Eurostar (UK) operates international passenger-only services between the United Kingdom and continental Europe, jointly with French National Railways (SNCF) and Belgian National Railways (SNCB/NMBS). Eurostar (UK) is a subsidiary of London & Continental Railways Ltd., which is jointly owned by National Express Group PLC and British Airways plc.

In addition, a service for the conveyance of accompanied road vehicles through the Channel Tunnel is provided by the tunnel operating company, Eurotunnel.

FREIGHT TRAIN OPERATIONS

Freight train services are operated under 'Open Access' arrangements by English Welsh & Scottish Railway (EWS), Freightliner, GB Railfreight Ltd.,Direct Rail Services and Mendip Rail

USING THIS BOOK

LAYOUT OF INFORMATION

DMUs are listed in numerical order of class, then in numerical order of set – using current numbers as allocated by the RSL. Individual 'loose' vehicles are listed in numerical order after vehicles formed into fixed formations. Where numbers carried are differ from those officially allocated these are noted in class headings where appropriate. Where sets or vehicles have been renumbered since the previous edition of this book, former numbering detail is shown in parentheses. Each entry is laid out as in the following example:

Set No.	Detail	Livery	Owner	Operation	Depot	Formation		Name
150 257	r*	**AR**	P	*AR*	NC	52257	57257	QUEEN BOADICEA

CLASS HEADINGS

Principal details and dimensions are quoted for each class in metric and/or imperial units as considered appropriate bearing in mind common usage in the UK.

The following abbreviations are used in class headings and also throughout this publication:

BR	British Railways.	kW	kilowatts.
BSI	Bergische Stahl Industrie.	lbf	pounds force.
DEMU	Diesel-electric multiple unit.	mm.	millimetres.
DMU	Diesel multiple unit (general term).	m.	metres.
GWR	Great Western Railway.	m.p.h.	miles per hour.
h.p.	horsepower.	r.p.m.	revolutions per minute.
Hz	Hertz.	RSL	Rolling Stock Library.
kN	kilonewtons.	t.	tonnes.
km/h	kilometres per hour.	V	volts.

All dimensions and weights are quoted for vehicles in an 'as new' condition with all necessary supplies (e.g. oil, water, sand) on board. Dimensions are quoted in the order Length – Width. All lengths quoted are over buffers or couplers as appropriate. All width dimensions quoted are maxima.

DETAIL DIFFERENCES

Only detail differences which currently affect the areas and types of train which vehicles may work are shown. All other detail differences are specifically excluded. Where such differences occur within a class, these are shown either in the heading information or alongside the individual set or vehicle number. The following standard abbreviation is used:

r   Radio Electronic Token Block (RETB) equipment.

In all cases use of the above abbreviations indicates the equipment indicated is normally operable. Meaning of non-standard abbreviations is detailed in individual class headings.

LIVERY CODES

Livery codes are used to denote the various liveries carried. Readers should note it is impossible in a publication of this size to list every livery variation which currently exists. In particular items ignored for the purposes of this book include minor colour variations, all numbering, lettering and branding and omission of logos.

The descriptions below are thus a general guide only and may be subject to slight variation between individual vehicles. Logos as appropriate for each livery are normally deemed to be carried. A complete list of livery codes used appears on page 78.

OWNER CODES

Owner codes are used in this to denote the owners of vehicles listed. Most vehicles are leased by the TOCs from specialist leasing companies. A complete list of owner codes used appears on page 79.

OPERATION CODES

Operation codes are used to denote the normal usage of the vehicles listed – i.e. A guide to the services of which train operating company any vehicle will normally be used upon. Where vehicles are used for non revenue earning purposes, an indication to the normal type of usage is given in the class heading. Where no operation code is shown, vehicles are currently not in use. A complete list of operation codes used appears on page 79.

DEPOT & LOCATION CODES

Depot codes are used to denote the normal maintenance base of each operational vehicle. However, maintenance may be carried out at other locations and may also be carried out by mobile maintenance teams.

Location codes are used to denote common storage locations whilst the full place name is used for other locations. A complete list of depot and location codes used appears on page 80.

SET FORMATIONS

Regular set formations are shown where these are normally maintained. Readers should note set formations might be temporarily varied from time to time to suit maintenance and/or operational requirements. Vehicles shown as 'Spare' are not formed in any regular set formation.

NAMES

Only names carried with official sanction are listed. As far as possible names are shown in UPPER/lower case characters as actually shown on the name carried on the vehicle(s). Unless otherwise shown, complete units are regarded as named rather than just the individual car(s) which carry the name.

GENERAL INFORMATION

CLASSIFICATION AND NUMBERING

First generation ('Heritage') DMUs are classified in the series 100–139.
Second generation DMUs are classified in the series 140–199.
Diesel-electric multiple units are classified in the series 200–249.
Service units are classified in the series 930–999.
First and second generation individual cars are numbered in the series 50000–59999 and 79000–79999.

DEMU individual cars are numbered in the series 60000–60999, except for a few former EMU vehicles which retain their EMU numbers.

Service stock individual cars are numbered in the series 975000–975999 and 977000–977999, although this series is not exclusively used for DMU vehicles.

OPERATING CODES

These codes are used by train operating company staff to describe the various different types of vehicles and normally appear on data panels on the inner (i.e. non driving) ends of vehicles.

DM	Driving Motor.	MF	Motor First
DMB	Driving Motor Brake.	MS	Motor Standard.
DMBS	Driving Motor Brake Standard	MSLRB	Motor Standard with buffet
DMC	Driving Motor Composite.	or MSRB	
DMF	Driving Motor First.	T	Trailer..
DMS	Driving Motor Standard.	TC	Trailer Composite.
DT	Driving Trailer.	TCso	Trailer Composite (semi-
DTC	Driving Trailer Composite.		open).
DTS	Driving Trailer Standard.	TS	Trailer Standard.
DTCso	Driving Trailer Composite	TSRB	Trailer Standard with buffet
			(semi-open).

All vehicles are of open configuration except where shown. A semi-open vehicle features both open and compartment accommodation, with first class accommodation usually in compartments in composite vehicles. Where two vehicles of the same type are formed within the same unit, the above codes may be suffixed by (A) and (B) to differentiate between the vehicles. The suffix 'L' denotes vehicles with a lavatory compartment.

A composite is a vehicle containing both first and standard class accommodation, whilst a brake vehicle is a vehicle containing separate specific accommodation for the conductor.

DESIGN CODES AND DIAGRAM CODES

For each type of vehicle the RSL issues a seven character 'Design Code' consisting of two letters plus four numbers and a suffix letter. (e.g. DP2010A). The first five characters of the Design Code are known as the 'Diagram Code' and these are quoted in this publication in sub-headings. The meaning of the various characters of the Design Code is as follows:

First Character

D   Diesel Multiple Unit vehicle.

Second Character

B   DEMU Driving motor passenger vehicle with brake compartment.
C   DEMU Driving motor passenger vehicle.
D   DEMU Non-driving motor passenger vehicle.
E   DEMU Driving trailer passenger vehicle.
F   DEMU Driving motor passenger vehicle (tilting).
G   DEMU Non-driving motor passenger vehicle (tilting).
H   DEMU Trailer passenger vehicle.
P   DMU (excl. DEMU) Driving motor passenger vehicle.
Q   DMU (excl. DEMU) Driving motor passenger vehicle with brake compartment.
R   DMU (excl. DEMU) Non-driving motor passenger vehicle.
S   DMU (excl. DEMU) Driving trailer passenger vehicle.
T   DMU (excl. DEMU) Trailer passenger vehicle.
X   DMU (excl. DEMU) Single unit railcar.
Z   All types of service vehicle.

Third Character

2   Standard class accommodation.
3   Composite accommodation.
5   No passenger accommodation.

Fourth & Fifth Characters

These distinguish between different designs of vehicle, each design being allocated a unique two digit number.

Special Note

Where vehicles have been declassified, the correct design code for a declassified vehicle is quoted in this publication, even though this may be at variance with RSL records, which do not always show the reality of the current position.

BUILD DETAILS

Lot Numbers

Vehicles ordered under the auspices of BR were allocated a Lot (batch) number when ordered and these are quoted in class headings and sub-headings.

ACCOMMODATION

The information given in class headings and sub-headings is in the form F/S nT (or TD) nW. For example 12/54 1T 1W denotes 12 first class and 54 standard class seats, 1 toilet and 1 wheelchair space. In declassified vehicles the capacity is still shown in terms of first and standard class seats whilst different types of seat remain fitted. TD denotes a toilet suitable for a disabled person.

1. DIESEL MECHANICAL & DIESEL HYDRAULIC UNITS

FIRST GENERATION UNITS

CLASS 101   METRO-CAMMELL

Class 101 are the last first generation DMUs left in passenger service and are scheduled for early withdrawal.
DMBS–DTSL–DMBS or DMBS–DMSL.
Construction: Aluminium alloy body on steel underframe.
Engines: Two Leyland 680/1 of 112 kW (150 h.p.) at 1800 r.p.m. per power car.
Transmission: Mechanical. Cardan shaft and freewheel to a four-speed epicyclic gearbox with a further cardan shaft to the final drive, each engine driving the inner axle of one bogie.
Brakes: Vacuum.
Gangways: British Standard (Midland scissors type). Within unit only.
Bogies: DD15 (motor) and DT11 (trailer).
Couplers: Screw couplings.
Dimensions: 18.49 x 2.82 m.
Seating Layout: 3+2 unidirectional (2+2 facing in first class).
Doors: Manually-operated slam.
Multiple Working: 'Blue Square' coupling code. First generation vehicles may be coupled together to work in multiple up to a maximum of 6 motor cars or 12 cars in total in a formation. First generation vehicles may not be coupled in multiple with second generation vehicles.
Maximum Speed: 70 m.p.h.

51175–51253. DMBS. Dia. DQ202. Lot No. 30467 1958–59. –/52. 32.5 t.
51426–51442. DMBS. Dia. DQ202. Lot No. 30500 1959. –/52. 32.5 t.
53204. DMBS. Dia. DQ202. Lot No. 30259 1957. –/52. 32.5 t.
53228. DMBS. Dia. DQ202. Lot No. 30261 1957. –/52. 32.5 t.
53253–53256. DMBS. Dia. DQ202. Lot No. 30266 1957. –/52. 32.5 t.
51496–51533. DMCL or DMSL. Dia. DP210. Lot No. 30501 1959. –/72 1T. 32.5t.
51803. DMSL. Dia. DP210. Lot No. 30588 1959. –/72 1T. 32.5 t.
53160–53163. DMSL. Dia. DP214. Lot No. 30253 1956. –/72 1T. 32.5 t.
53170–53171. DMSL. Dia. DP214. Lot No. 30255 1957. –/72 1T. 32.5 t.
53266–53269. DMSL. Dia. DP210. Lot No. 30267 1957. –/72 1T. 32.5 t.
53746. DMSL. Dia. DP210. Lot No. 30271 1957. –/72 1T. 32.5 t.
54056. DTSL. Dia. DS206. Lot No. 30260 1957. –/72 1T. 25.5 t.
54347–54408. DTSL. Dia. DS206. Lot No. 30468 1958. –/72 1T. 25.5 t.

3-car Sets. DMBS–DTSL plus DMBS locked out of use.

101 653	**RR**	A	*NW*	LO	51426	54358	53228
101 654	**RR**	A	*NW*	LO	51800	54408	51175
101 656	**RR**	A	*NW*	LO	51230	54056	51428
101 663	**RR**	A	*NW*	LO	51201	54347	51442

Twin Power Car Sets. DMBS–DMSL.

Non-Standard livery: Caledonian style blue with yellow/orange stripes.

101 676	**RR**	A	*NW*	LO	51205	51803
101 677	**RR**	A	*NW*	LO	51179	51496
101 678	**RR**	A	*NW*	LO	51210	53746
101 679	**RR**	A	*NW*	LO	51224	51533
101 680	**RR**	A	*NW*	LO	53204	53163
101 681	**RR**	A	*NW*	LO	51228	51506
101 682	**RR**	A	*NW*	LO	53256	51505
101 683	**RR**	A	*NW*	LO	51177	53269
101 684	**S**	A	*NW*	LO	51187	51509
101 685	**G**	A	*NW*	LO	51164	53160
101 687	**S**	A	*NW*	LO	51247	51512
101 689	**S**	A	*NW*	LO	51185	51511
101 691	**S**	A	*NW*	LO	51253	53171
101 692	**0**	A	*NW*	LO	53253	53170
101 693	**S**	A	*NW*	LO	51192	53266
101 694	**S**	A	*NW*	LO	51188	53268
101 695	**S**	A	*NW*	LO	51226	51499

SECOND GENERATION UNITS

All second generation units in this section have air brakes and are equipped with public address, with transmission equipment on driving vehicles and flexible diaphragm gangways. Except where otherwise stated, transmission is Voith 211r hydraulic with a cardan shaft to a Gmeinder GM190 final drive.

CLASS 142   PACER   BREL/LEYLAND

DMS–DMSL.

Construction: Steel. Built from Leyland National bus parts on four-wheeled underframes.
Engines: One Cummins LTA10-R of 172 kW (230 h.p.) at 2100 r.p.m. (* One Perkins 2006-TWH of 172 kW (230 h.p.) at 2100 r.p.m.) per car.
Couplers: BSI at outer ends, bar within unit.
Seating Layout: 3+2 mainly unidirectional bus style.
Dimensions: 15.66 x 2.80 m.
Gangways: Within unit only.   **Wheel Arrangement:** 1-A A-1.
Doors: Twin-leaf inward pivoting.   **Maximum Speed:** 75 m.p.h.
Multiple Working: Classes 142, 143, 144, 150, 153, 155, 156, 158, 159, 170.

55542–55591. DMS. Dia. DP234 (s DP271). Lot No. 31003 BREL Derby 1985–86. –/62. (s –/56, t –/55, u –/64) 23.26 t.
55592–55641. DMSL. Dia. DP235 (s DP272). Lot No. 31004 BREL Derby 1985–86. –/59 1T. (s –/50, u –/54) 24.97 t.
55701–55746. DMS. Dia. DP234 (s DP271). Lot No. 31013 BREL Derby 1986–87. –/62. (s –/56, t –/55, u –/64) 23.26 t.
55747–55792. DMSL. Dia. DP235 (s DP272). Lot No. 31014 BREL Derby 1986–87. –/59 1T. (s –/50, u –/54) 24.97 t.

s Fitted with 2+2 individual high-backed seating.
t DMS fitted with luggage rack, seating –/55.
u Fitted with 3+2 individual low-back seating.

142 001	t	**GM**	A	*NW*	NH	55542 55592
142 002		**GM**	A	*NW*	NH	55543 55593
142 003		**GM**	A	*NW*	NH	55544 55594
142 004	t	**GM**	A	*NW*	NH	55545 55595
142 005	t	**GM**	A	*NW*	NH	55546 55596
142 006		**GM**	A	*NW*	NH	55547 55597
142 007	t	**GM**	A	*NW*	NH	55548 55598
142 008	t	**GM**	A		ZC (S)	55549 55599
142 009	t	**GM**	A	*NW*	NH	55550 55600
142 010		**GM**	A	*NW*	NH	55551 55601
142 011	t	**GM**	A	*NW*	NH	55552 55602
142 012	t	**GM**	A	*NW*	NH	55553 55603
142 013		**GM**	A	*NW*	NH	55554 55604
142 014	t	**GM**	A	*NW*	NH	55555 55605
142 015	s	**RR**	A	*NS*	HT	55556 55606
142 016	s	**RR**	A	*NS*	HT	55557 55607
142 017	s	**TW**	A	*NS*	HT	55558 55608
142 018	s	**TW**	A	*NS*	HT	55559 55609
142 019	s	**TW**	A	*NS*	HT	55560 55610
142 020	s	**TW**	A	*NS*	HT	55561 55611
142 021	s	**TW**	A	*NS*	HT	55562 55612
142 022	s	**TW**	A	*NS*	HT	55563 55613
142 023	t	**NW**	A	*NW*	NH	55564 55614
142 024	s	**RR**	A	*NS*	HT	55565 55615
142 025	s	**NS**	A	*NS*	HT	55566 55616
142 026	s	**NS**	A	*NS*	HT	55567 55617
142 027	t	**GM**	A	*NW*	NH	55568 55618
142 028	t	**GM**	A	*NW*	NH	55569 55619
142 029		**GM**	A	*NW*	NH	55570 55620
142 030		**GM**	A	*NW*	NH	55571 55621
142 031	t	**GM**	A	*NW*	NH	55572 55622
142 032	t	**GM**	A	*NW*	NH	55573 55623
142 033	t	**RR**	A	*NW*	NH	55574 55624
142 034	t	**GM**	A	*NW*	NH	55575 55625
142 035	t	**GM**	A	*NW*	NH	55576 55626
142 036	t	**RR**	A	*NW*	NH	55577 55627
142 037	t	**GM**	A	*NW*	NH	55578 55628
142 038	t	**GM**	A	*NW*	NH	55579 55629
142 039	t	**GM**	A	*NW*	NH	55580 55630
142 040	t	**GM**	A	*NW*	NH	55581 55631
142 041	u	**MY**	A	*NW*	NH	55582 55632
142 042	u	**MY**	A	*NW*	NH	55583 55633
142 043		**GM**	A	*NW*	NH	55584 55634
142 044	u	**MY**	A	*NW*	NH	55585 55635
142 045		**GM**	A	*NW*	NH	55586 55636
142 046		**GM**	A	*NW*	NH	55587 55637
142 047		**RR**	A	*NW*	NH	55588 55638

142 048		**RR**	A	*NW*	NH	55589 55639
142 049		**GM**	A	*NW*	NH	55590 55640
142 050	s	**NS**	A	*NS*	HT	55591 55641
142 051	u	**MT**	A	*NW*	NH	55701 55747
142 052	u	**MT**	A	*NW*	NH	55702 55748
142 053	u	**MT**	A	*NW*	NH	55703 55749
142 054	u	**MT**	A	*NW*	NH	55704 55750
142 055	u	**MT**	A	*NW*	NH	55705 55751
142 056	u	**MT**	A	*NW*	NH	55706 55752
142 057	u	**MT**	A	*NW*	NH	55707 55753
142 058	u	**MT**	A	*NW*	NH	55708 55754
142 060	t	**GM**	A	*NW*	NH	55710 55756
142 061		**GM**	A	*NW*	NH	55711 55757
142 062	t	**GM**	A	*NW*	NH	55712 55758
142 063	t	**GM**	A	*NW*	NH	55713 55759
142 064	t	**GM**	A	*NW*	NH	55714 55760
142 065	s	**NS**	A	*NS*	HT	55715 55761
142 066	s	**NS**	A	*NS*	NL	55716 55762
142 067		**GM**	A	*NW*	NH	55717 55763
142 068	t	**GM**	A	*NW*	NH	55718 55764
142 069		**GM**	A	*NW*	NH	55719 55765
142 070	t	**GM**	A	*NW*	NH	55720 55766
142 071	s	**RR**	A	*NS*	HT	55721 55767
142 072		**RR**	A	*NS*	NL	55722 55768
142 073		**RR**	A	*NS*	NL	55723 55769
142 074		**RR**	A	*NS*	NL	55724 55770
142 075		**RR**	A	*NS*	NL	55725 55771
142 076		**RR**	A	*NS*	NL	55726 55772
142 077		**RR**	A	*NS*	NL	55727 55773
142 078	s	**RR**	A	*NS*	NL	55728 55774
142 079		**RR**	A	*NS*	NL	55729 55775
142 080		**RR**	A	*NS*	NL	55730 55776
142 081		**RR**	A	*NS*	NL	55731 55777
142 082		**RR**	A	*NS*	NL	55732 55778
142 083		**RR**	A	*CA*	CF	55733 55779
142 084	s*	**RR**	A	*NS*	NL	55734 55780
142 085	s	**RR**	A	*CA*	CF	55735 55781
142 086	s	**RR**	A	*CA*	CF	55736 55782
142 087	s	**RR**	A	*CA*	CF	55737 55783
142 088	s	**RR**	A	*CA*	CF	55738 55784
142 089	s	**RR**	A	*CA*	CF	55739 55785
142 090	s	**RR**	A	*CA*	CF	55740 55786
142 091	s	**RR**	A	*CA*	CF	55741 55787
142 092	s	**RR**	A	*CA*	CF	55742 55788
142 093	s	**RR**	A	*CA*	CF	55743 55789
142 094	s	**RR**	A	*CA*	CF	55744 55790
142 095	s	**RR**	A	*NS*	NL	55745 55791
142 096	s	**RR**	A	*CA*	CF	55746 55792

CLASS 143   PACER   ALEXANDER/BARCLAY

DMS–DMSL. Similar design to Class 142, but bodies built by W. Alexander with Barclay underframes.

Construction: Steel. Alexander bus bodywork on four-wheeled underframes.
Engines: One Cummins LTA10-R of 172 kW (230 h.p.) at 2100 r.p.m. per car.
Couplers: BSI at outer ends, bar couplers within unit.
Seating Layout: 3+2 mainly unidirectional bus style.
Dimensions: 15.55 x 2.70 m.
Gangways: Within unit only.   **Wheel Arrangement:** 1-A A-1.
Doors: Twin-leaf inward pivoting.   **Maximum Speed:** 75 m.p.h.
Multiple Working: Classes 142, 143, 144, 150, 153, 155, 156, 158, 159, 170.

DMS. Dia. DP236 Lot No. 31005 Andrew Barclay 1985–86. –/62 (s –/55). 24.5 t.
DMSL. Dia. DP237 Lot No. 31006 Andrew Barclay 1985–86. –/60 1T (s –/51 1T). 25.0 t.

s Fitted with 2+2 individual high-backed seating.

143 601		**RR**	RD	*WW*	CF	55642	55667	
143 602		**RR**	P	*CA*	CF	55651	55668	
143 603		**RR**	P	*CA*	CF	55658	55669	
143 604		**RR**	P	*CA*	CF	55645	55670	
143 605		**RR**	P	*CA*	CF	55646	55671	Crimestoppers
143 606	s	**VL**	P	*CA*	CF	55647	55672	
143 607		**RR**	P	*CA*	CF	55648	55673	
143 608		**RR**	P	*CA*	CF	55649	55674	
143 609		**RR**	BB	*CA*	CF	55650	55675	TOM JONES
143 610		**RR**	RD	*WW*	CF	55643	55676	
143 611		**AL**	P	*CA*	CF	55652	55677	
143 612		**RR**	P	*WW*	CF	55653	55678	
143 613		**AL**	P	*CA*	CF	55654	55679	
143 614		**RR**	RD	*WW*	CF	55655	55680	
143 615		**RR**	P	*CA*	CF	55656	55681	
143 616		**RR**	P	*CA*	CF	55657	55682	
143 617		**RR**	RI	*WW*	CF	55644	55683	
143 618		**RR**	RI	*WW*	CF	55659	55684	
143 619		**RR**	RI	*WW*	CF	55660	55685	
143 620		**RR**	P	*WW*	CF	55661	55686	
143 621		**RR**	P	*WW*	CF	55662	55687	
143 622		**RR**	P	*WW*	CF	55663	55688	
143 623		**RR**	P	*WW*	CF	55664	55689	
143 624		**RR**	P	*CA*	CF	55665	55690	
143 625		**RR**	P	*CA*	CF	55666	55691	

CLASS 144   PACER   ALEXANDER/BREL

DMS–DMSL or DMS–MS–DMSL. As Class 143, but underframes built by BREL.

Construction: Steel. Alexander bus bodywork on four-wheeled underframes.

Engines: One Cummins LTA10-R of 172 kW (230 h.p.) at 2100 r.p.m. per car.
Couplers: BSI at outer ends, bar couplers within unit.
Seating Layout: 3+2 mainly unidirectional bus style.
Dimensions: 15.11 x 2.73 m.
Gangways: Within unit only.   **Wheel Arrangement:** 1-A A-1.
Doors: Twin-leaf inward pivoting.   **Maximum Speed:** 75 m.p.h.
Multiple Working: Classes 142, 143, 144, 150, 153, 155, 156, 158, 159, 170.

DMS. Dia. DP240 Lot No. 31015 BREL Derby 1986–87. –/62 1W. 24.2 t.
MS. Dia. DR205 Lot No. BREL Derby 31037 1987. –/73. 22.6 t.
DMSL. Dia. DP241 Lot No. BREL Derby 31016 1986–87. –/60 1T. 25.0 t.

Note: The centre cars of the 3-car units are owned by West Yorkshire PTE, although managed by Porterbrook Leasing Company.

144 001	**WY**	P	*NS*	NL	55801		55824
144 002	**WY**	P	*NS*	NL	55802		55825
144 003	**WY**	P	*NS*	NL	55803		55826
144 004	**WY**	P	*NS*	NL	55804		55827
144 005	**WY**	P	*NS*	NL	55805		55828
144 006	**WY**	P	*NS*	NL	55806		55829
144 007	**WY**	P	*NS*	NL	55807		55830
144 008	**WY**	P	*NS*	NL	55808		55831
144 009	**WY**	P	*NS*	NL	55809		55832
144 010	**WY**	P	*NS*	NL	55810		55833
144 011	**RR**	P	*NS*	NL	55811		55834
144 012	**RR**	P	*NS*	NL	55812		55835
144 013	**RR**	P	*NS*	NL	55813		55836
144 014	**WY**	P	*NS*	NL	55814	55850	55837
144 015	**WY**	P	*NS*	NL	55815	55851	55838
144 016	**WY**	P	*NS*	NL	55816	55852	55839
144 017	**WY**	P	*NS*	NL	55817	55853	55840
144 018	**WY**	P	*NS*	NL	55818	55854	55841
144 019	**WY**	P	*NS*	NL	55819	55855	55842
144 020	**WY**	P	*NS*	NL	55820	55856	55843
144 021	**WY**	P	*NS*	NL	55821	55857	55844
144 022	**WY**	P	*NS*	NL	55822	55858	55845
144 023	**WY**	P	*NS*	NL	55823	55859	55846

CLASS 150/0   SPRINTER   BREL

DMSL–MS–DMS. Prototype Sprinter.

Construction: Steel.
Engines: One Cummins NT-855-R4 of 213 kW (285 h.p.) at 2100 r.p.m. per car.
Bogies: BX8P (powered), BX8T (non-powered).
Couplers: BSI at outer end of driving vehicles, bar non-driving ends.
Seating Layout: 3+2 (mainly unidirectional).
Dimensions: 20.06 x 2.82 m (outer cars), 20.18 x 2.82 m (inner car).
Gangways: Within unit only.   **Wheel Arrangement:** 2-B – 2-B – B-2.
Doors: Single-leaf sliding.   **Maximum Speed:** 75 m.p.h.
Multiple Working: Classes 142, 143, 144, 150, 153, 155, 156, 158, 159, 170.

DMSL. Dia. DP230. Lot No. 30984 BREL York 1984. –/72 1T. 35.8 t.
MS. Dia. DR202. Lot No. 30986 BREL York 1984. –/92. 34.4 t.
DMS. Dia. DP231. Lot No. 30985 BREL York 1984. –/76. 35.6 t.

150 001	r	**CO**	A	*CT*	TS	52200	55400	55300
150 002	r	**CO**	A	*CT*	TS	52201	55401	55301

CLASS 150/1   SPRINTER   BREL

DMSL–DMS or DMSL–DMSL–DMS or DMSL–DMS–DMS.

Construction: Steel.
Engines: One Cummins NT855R5 of 213 kW (285 h.p.) at 2100 r.p.m. per car.
Bogies: BP38 (powered), BT38 (non-powered).
Couplers: BSI.
Seating Layout: 3+2 facing as built but 150010–150 132 were reseated with mainly unidirectional seating.
Dimensions: 19.74 x 2.82 m.
Gangways: Within unit only.   **Wheel Arrangement:** 2-B (– 2-B) – B-2.
Doors: Single-leaf sliding.   **Maximum Speed:** 75 m.p.h.
Multiple Working: Classes 142, 143, 144, 150, 153, 155, 156, 158, 159, 170.

DMSL. Dia. DP238. Lot No. 31011 BREL York 1985–86. –/72 1T (s –/58 1TD, t –/71 1W 1T, u –/71 1T). 36.5 t.
DMS. Dia. DP239. Lot No. 31012 BREL York 1985–86. –/76 (s –/64). 38.45 t.

Notes: The centre cars of three-car units are Class 150/2 vehicles. For details see Class 150/2.
Units reliveried in **NW** livery have been refurbished with new seating.

150 010	ru	**CO**	A	*CT*	TS	52110	57226	57110
150 011	ru	**CO**	A	*CT*	TS	52111	52204	57111
150 012	ru	**CO**	A	*CT*	TS	52112	57206	57112
150 013	ru	**CO**	A	*CT*	TS	52113	52226	57113
150 014	ru	**CO**	A	*CT*	TS	52114	57204	57114
150 015	ru	**CO**	A	*CT*	TS	52115	52206	57115
150 016	ru	**CO**	A	*CT*	TS	52116	57212	57116
150 018	r	**CO**	A	*CT*	TS	52118	52220	57118
150 019	ru	**CO**	A	*CT*	TS	52119	57220	57119
150 101	ru	**CO**	A	*CT*	TS	52101		57101
150 102	ru	**CO**	A	*CT*	TS	52102		57102
150 103	ru	**CO**	A	*CT*	TS	52103		57103
150 104	r	**CO**	A	*CT*	TS	52104		57104
150 105	ru	**CO**	A	*CT*	TS	52105		57105
150 106	r	**CO**	A	*CT*	TS	52106		57106
150 107	r	**CO**	A	*CT*	TS	52107		57107
150 108	ru	**CO**	A	*CT*	TS	52108		57108
150 109	ru	**CO**	A	*CT*	TS	52109		57109
150 117	ru	**CO**	A	*CT*	TS	52117	57117	
150 120	t	**SL**	A	*SL*	BY	52120	57120	
150 121	r	**CO**	A	*CT*	TS	52121	57121	
150 122	ru	**CO**	A	*CT*	TS	52122	57122	
150 123		**CO**	A	*SL*	BY	52123	57123	

▲ Regional Railways liveried Class 101 No. 101 653 passes the site of the former Heeley Carriage Sidings whilst working the 16.15 Sheffield–Manchester Piccadilly service on 27th May 2000. **Wolfram Stein**

▼ Merseytravel liveried Class 142 No. 142 052 is far from home as it passes through Carlisle with a southbound empty stock working. The date is 25th June 2000. **K. Conkey**

▲ Carrying the new Cardiff Valley Lines livery, Class 143 No. 143 606 is pictured at Cardiff Central on 9th November 2000 whilst working the 11.02 Coryton–Maesteg. **Bob Sweet**

▼ 3-car Class 144 No. 144 014, in West Yorkshire PTE livery, passes Gascoigne Wood with a Leeds–Hull train on 30th March 2000. **Ian A. Lyall**

▲ Class 117 and 121 DMUs have now been withdrawn from Bedford–Bletchley services in favour of Class 150/1s. Class 150/1 No. 150 131, in Silverlink livery, is seen here at Bletchley on 5th August 2000 shortly before working the 19.22 service to Bedford. **Martyn Hilbert**

▼ The North Western Trains livery, albeit with First North Western branding, is now being applied to Class 150/2s on refurbishment. 150 207 is pictured leaving Leyland with the 17.05 Blackpool North–Manchester Piccadilly on 19th August 2000. **Martyn Hilbert**

▲ A pair of Class 153s, Nos. 153 332 and 153 330, enter Lancaster on 30th September 2000 whilst forming the 15.00 Barrow-in-Furness–Manchester Airport.
Dave McAlone

▼ West Yorkshire PTE liveried Class 155 No. 155 344 approaches Manston Crossing with the 09.23 Manchester Victoria–Selby on 5th October 1999.
John G. Teasdale

Northern Spirit liveried Class 156 No. 156 448 is pictured south of Armathwaite on 27th July 2000 with the 16.48 Carlisle–Leeds.

K. Conkey

▲ Class 156 No. 156 449, in Scotrail livery, passes Enterkinfoot, north of Dumfries, on 26th July 2000 with the 13.07 Carlisle–Glasgow Central. **K. Conkey**

▼ Northern Spirit units which are used on Transpennine Express services have a distinctive colour scheme applied. One of the units which carries the livery, 3-car Class 158 No. 158 799, is seen passing through Horbury Cutting on 16th October 2000. **G.W. Morrison**

Wales & West 'Alphaline' liveried Class 158 No. 158 746 is about to enter Parsons Tunnel, near Teignmouth, whilst forming the 11.45 Penzance–Cardiff Central service on 29th April 2000.

John Chalcraft

A pair of Class 159 units Nos. 159 007, in South West Trains livery, and 159 017, in the old Network SouthEast livery, are pictured forming the 12.35 London Waterloo–Exeter St Davids service as they pull away from Clapham Junction. The date is 20th April 2000. **K. Conkey**

150 124	ru	**CO**	A	*CT*	TS	52124	57124
150 125	ru	**CO**	A	*CT*	TS	52125	57125
150 126	ru	**CO**	A	*CT*	TS	52126	57126
150 127	t	**SL**	A	*SL*	BY	52127	57127	Bletchley TMD
150 128		**CO**	A	*SL*	BY	52128	57128
150 129	t	**SL**	A	*SL*	BY	52129	57129	MARSTON VALE
150 130	t	**SL**	A	*SL*	BY	52130	57130	LESLIE CRABBE
150 131	t	**SL**	A	*SL*	BY	52131	57131
150 132	r	**CO**	A	*CT*	TS	52132	57132
150 133	s	**NW**	A	*NW*	NH	52133	57133
150 134	s	**NW**	A	*NW*	NH	52134	57134
150 135	s	**NW**	A	*NW*	NH	52135	57135
150 136	s	**NW**	A	*NW*	NH	52136	57136
150 137	s	**NW**	A	*NW*	NH	52137	57137
150 138	s	**NW**	A	*NW*	NH	52138	57138
150 139	s	**NW**	A	*NW*	NH	52139	57139
150 140	s	**NW**	A	*NW*	NH	52140	57140
150 141	s	**NW**	A	*NW*	NH	52141	57141
150 142	s	**NW**	A	*NW*	NH	52142	57142
150 143	s	**NW**	A	*NW*	NH	52143	57143
150 144	s	**NW**	A	*NW*	NH	52144	57144
150 145	s	**NW**	A	*NW*	NH	52145	57145
150 146	s	**NW**	A	*NW*	NH	52146	57146
150 147	s	**NW**	A	*NW*	NH	52147	57147
150 148	s	**NW**	A	*NW*	NH	52148	57148
150 149	s	**NW**	A	*NW*	NH	52149	57149
150 150	s	**NW**	A	*NW*	NH	52150	57150

CLASS 150/2   SPRINTER   BREL

DMSL–DMS.

Construction: Steel.
Engines: One Cummins NT855R5 of 213 kW (285 h.p.) at 2100 r.p.m. per car.
Bogies: BP38 (powered), BT38 (non-powered).
Couplers: BSI.
Seating Layout: 3+2 mainly unidirectional seating.
Dimensions: 19.74 x 2.82 m.
Gangways: Throughout.   **Wheel Arrangement:** 2-B – B-2.
Doors: Single-leaf sliding.   **Maximum Speed:** 75 m.p.h.
Multiple Working: Classes 142, 143, 144, 150, 153, 155, 156, 158, 159, 170.

DMSL. Dia. DP242. Lot No. 31017 BREL York 1986–87. –/73 1T (s –/70 1TD, t /70 1T). 35.8 t.
DMS. Dia. DP243. Lot No. 31018 BREL York 1986–87. –/76 (* –/68, s –/62, u –/73). 34.9 t.

Units in **NW** livery have been refurbished with new seating.

150 201	s	**NW**	A	*NW*	NH	52201	57201
150 202		**CO**	A	*CT*	TS	52202	57202
150 203	s	**NW**	A	*NW*	NH	52203	57203
150 205	s	**NW**	A	*NW*	NH	52205	57205

150 207	s	**NW**	A	*NW*	NH	52207	57207	
150 208		**RR**	P	*SR*	HA	52208	57208	
150 210		**CO**	A	*CT*	TS	52210	57210	
150 211	s	**NW**	A	*NW*	NH	52211	57211	
150 213	r*	**PS**	P	*AR*	NC	52213	57213	LORD NELSON
150 214		**CO**	A	*CT*	TS	52214	57214	
150 215	s	**NW**	A	*NW*	NH	52215	57215	
150 216		**CO**	A	*CT*	TS	52216	57216	
150 217	r*	**PS**	P	*AR*	NC	52217	57217	OLIVER CROMWELL
150 218	s	**NW**	A	*NW*	NH	52218	57218	
150 219	r	**RR**	P	*WW*	CF	52219	57219	
150 221	r	**RR**	P	*WW*	CF	52221	57221	
150 222	s	**NW**	A	*NW*	NH	52222	57222	
150 223	s	**NW**	A	*NW*	NH	52223	57223	
150 224	t	**GM**	A	*NW*	NH	52224	57224	
150 225	u	**GM**	A	*NW*	NH	52225	57225	
150 227	r*	**PS**	P	*AR*	NC	52227	57227	SIR ALF RAMSEY
150 228	tu	**RR**	P	*NS*	NL	52228	57228	
150 229	r*	**PS**	P	*AR*	NC	52229	57229	GEORGE BORROW
150 230	r	**RR**	P	*WW*	CF	52230	57230	
150 231	r*	**PS**	P	*AR*	NC	52231	57231	KING EDMUND
150 232	r	**RR**	P	*WW*	CF	52232	57232	
150 233	r	**RR**	P	*WW*	CF	52233	57233	
150 234	r	**RR**	P	*WW*	CF	52234	57234	
150 235	r*	**PS**	P	*AR*	NC	52235	57235	CARDINAL WOLSEY
150 236	r	**RR**	P	*CA*	CF	52236	57236	
150 237	r*	**PS**	P	*AR*	NC	52237	57237	HEREWARD THE WAKE
150 238	r	**RR**	P	*WW*	CF	52238	57238	
150 239	r	**RR**	P	*WW*	CF	52239	57239	
150 240	r	**RR**	P	*WW*	CF	52240	57240	
150 241	r	**RR**	P	*WW*	CF	52241	57241	
150 242	r	**RR**	P	*WW*	CF	52242	57242	
150 243	r	**RR**	P	*WW*	CF	52243	57243	
150 244	r	**RR**	P	*WW*	CF	52244	57244	
150 245	r	**RR**	P	*NS*	NL	52245	57245	
150 246	r	**RR**	P	*WW*	CF	52246	57246	
150 247	r	**RR**	P	*WW*	CF	52247	57247	
150 248	r	**RR**	P	*WW*	CF	52248	57248	
150 249	r	**RR**	P	*WW*	CF	52249	57249	
150 250		**RR**	P	*SR*	HA	52250	57250	
150 251	r	**RR**	P	*WW*	CF	52251	57251	
150 252		**RR**	P	*SR*	HA	52252	57252	
150 253	r	**RR**	P	*WW*	CF	52253	57253	
150 254	r	**RR**	P	*WW*	CF	52254	57254	
150 255	r*	**AR**	P	*AR*	NC	52255	57255	HENRY BLOGG
150 256		**RR**	P	*SR*	HA	52256	57256	
150 257	r*	**AR**	P	*AR*	NC	52257	57257	QUEEN BOADICEA
150 258		**RR**	P	*SR*	HA	52258	57258	
150 259		**RR**	P	*SR*	HA	52259	57259	
150 260		**RR**	P	*SR*	HA	52260	57260	
150 261	r	**RR**	P	*WW*	CF	52261	57261	

150 262		**RR**	P	*SR*	HA	52262	57262
150 263	r	**RR**	P	*WW*	CF	52263	57263
150 264		**RR**	P	*SR*	HA	52264	57264
150 265	r	**RR**	P	*WW*	CF	52265	57265
150 266	r	**RR**	P	*WW*	CF	52266	57266
150 267	r	**RR**	P	*WW*	CF	52267	57267
150 268		**RR**	P	*NS*	NL	52268	57268
150 269		**RR**	P	*NS*	NL	52269	57269
150 270		**RR**	P	*NS*	NL	52270	57270
150 271		**RR**	P	*NS*	NL	52271	57271
150 272		**RR**	P	*NS*	NL	52272	57272
150 273		**RR**	P	*NS*	NL	52273	57273
150 274		**RR**	P	*NS*	NL	52274	57274
150 275		**RR**	P	*NS*	NL	52275	57275
150 276		**RR**	P	*NS*	NL	52276	57276
150 277		**RR**	P	*NS*	NL	52277	57277
150 278	r	**RR**	P	*WW*	CF	52278	57278
150 279		**RR**	P	*WW*	CF	52279	57279
150 280		**RR**	P	*CA*	CF	52280	57280
150 281		**RR**	P	*CA*	CF	52281	57281
150 282		**RR**	P	*CA*	CF	52282	57282
150 283		**RR**	P	*SR*	HA	52283	57283
150 284		**RR**	P	*SR*	HA	52284	57284
150 285		**RR**	P	*SR*	HA	52285	57285	EDINBURGH–BATHGATE 1986–1996
Spare		**CO**	A	*CT*	TS		57209

CLASS 153   SUPER SPRINTER   LEYLAND BUS

DMSL. Converted by Hunslet-Barclay, Kilmarnock from Class 155 two-car units.

Construction: Steel. Built from Leyland National bus parts on bogied underframes.
Engine: One Cummins NT855R5 of 213 kW (285 h.p.) at 2100 r.p.m.
Bogies: One P3-10 (powered) and one BT38 (non-powered).
Couplers: BSI.
Seating Layout: 2+2 facing/unidirectional.
Dimensions: 23.21 x 2.70 m.
Gangways: Throughout.   **Wheel Arrangement:** 2-B.
Doors: Single-leaf sliding plug.   **Maximum Speed:** 75 m.p.h.
Multiple Working: Classes 142, 143, 144, 150, 153, 155, 156, 158, 159, 170.

52301–52335. DMSL. Dia. DX203. Lot No. 31026 1987–88. Converted under Lot No. 31115 1991–2. –/72 1TD 1W (* –/66 1TD 1W). 41.2 t.
57301–57335. DMSL. Dia. DX203. Lot No. 31027 1987–88. Converted under Lot No. 31115 1991–2. –/72 1TD (* –/66 1TD). 41.2 t.

Notes:

Cars numbered in the 573XX series were renumbered by adding 50 to their original number so that the last two digits correspond with the set number.
Central Trains units have been fitted with new seating.
Wales & West units have been reseated with seats removed from that company's Class 158 units.

153 301		**RR**	A	*NS*	NL	52301	
153 302	r	**RR**	A	*WW*	CF	52302	
153 303	r	**RR**	A	*WW*	CF	52303	
153 304		**RR**	A	*NS*	NL	52304	
153 305	r	**RR**	A	*WW*	CF	52305	
153 306	r*	**PS**	P	*AR*	NC	52306	EDITH CAVELL
153 307		**RR**	A	*NS*	NL	52307	
153 308	r	**RR**	A	*WW*	CF	52308	
153 309	r*	**PS**	P	*AR*	NC	52309	GERARD FIENNES
153 310		**NW**	P	*NW*	NH	52310	
153 311	r*	**PS**	P	*AR*	NC	52311	JOHN CONSTABLE
153 312	r	**RR**	A	*WW*	CF	52312	
153 313		**NW**	P	*NW*	NH	52313	
153 314	r*	**PS**	P	*AR*	NC	52314	DELIA SMITH
153 315		**RR**	A	*NS*	NL	52315	
153 316		**NW**	P	*NW*	NH	52316	
153 317		**RR**	A	*NS*	NL	52317	
153 318	r	**RR**	A	*WW*	CF	52318	
153 319		**RR**	A	*NS*	NL	52319	
153 320		**RR**	P	*CT*	TS	52320	
153 321	r	**PS**	P	*CT*	TS	52321	
153 322	r*	**RR**	P	*AR*	NC	52322	BENJAMIN BRITTEN
153 323	r	**RR**	P	*CT*	TS	52323	
153 324		**NW**	P	*NW*	NH	52324	
153 325	r	**RR**	P	*CT*	TS	52325	
153 326	r*	**PS**	P	*AR*	NC	52326	TED ELLIS
153 327	r	**RR**	A	*WW*	CF	52327	
153 328		**RR**	A	*NS*	NL	52328	
153 329	r	**RR**	P	*CT*	TS	52329	
153 330		**NW**	P	*NW*	NH	52330	
153 331		**RR**	A	*NS*	NL	52331	
153 332		**NW**	P	*NW*	NH	52332	
153 333	r	**RR**	P	*CT*	TS	52333	
153 334	r	**RR**	P	*CT*	TS	52334	
153 335	r*	**PS**	P	*AR*	NC	52335	MICHAEL PALIN
153 351		**RR**	A	*NS*	NL	57351	
153 352		**RR**	A	*NS*	NL	57352	
153 353	r	**RR**	A	*WW*	CF	57353	
153 354	r	**RR**	P	*CT*	TS	57354	
153 355	r	**RR**	A	*WW*	CF	57355	
153 356	r	**RR**	P	*CT*	TS	57356	
153 357		**RR**	A	*NS*	NL	57357	
153 358		**NW**	P	*NW*	NH	57358	
153 359		**NW**	P	*NW*	NH	57359	
153 360		**RR**	P	*NW*	NH	57360	
153 361		**RR**	P	*NW*	NH	57361	
153 362	r	**RR**	A	*WW*	CF	57362	
153 363		**RR**	P	*NW*	NH	57363	
153 364	r	**RR**	P	*CT*	TS	57364	
153 365	r	**RR**	P	*CT*	TS	57365	

153 366	r	**RR**	P	*CT*	TS	57366
153 367		**RR**	P	*NW*	NH	57367
153 368	r	**RR**	A	*WW*	CF	57368
153 369	r	**RR**	P	*CT*	TS	57369
153 370	r	**RR**	A	*WW*	CF	57370
153 371	r	**RR**	P	*CT*	TS	57371
153 372	r	**RR**	A	*WW*	CF	57372
153 373	r	**RR**	A	*WW*	CF	57373
153 374	r	**RR**	A	*WW*	CF	57374
153 375	r	**RR**	P	*CT*	TS	57375
153 376	r	**RR**	P	*CT*	TS	57376
153 377	r	**RR**	A	*WW*	CF	57377
153 378		**RR**	A	*NS*	NL	57378
153 379	r	**RR**	P	*CT*	TS	57379
153 380	r	**RR**	A	*WW*	CF	57380
153 381	r	**RR**	P	*CT*	TS	57381
153 382	r	**RR**	A	*WW*	CF	57382
153 383	r	**RR**	P	*CT*	TS	57383
153 384	r	**RR**	P	*CT*	TS	57384
153 385	r	**RR**	P	*CT*	TS	57385

CLASS 155   SUPER SPRINTER   LEYLAND BUS

DMSL–DMS.

Construction: Steel. Built from Leyland National bus parts on bogied underframes.
Engines: One Cummins NT855R5 of 213 kW (285 h.p.) at 2100 r.p.m. per car
Bogies: One P3-10 (powered) and one BT38 (non-powered).
Couplers: BSI.
Seating Layout: 2+2 facing/unidirectional.
Dimensions: 23.21 x 2.70 m.
Gangways: Throughout.   **Wheel Arrangement:** 2-B – B-2.
Doors: Single-leaf sliding plug.   **Maximum Speed:** 75 m.p.h.
Multiple Working: Classes 142, 143, 144, 150, 153, 155, 156, 158, 159, 170.

DMSL. Dia. DP248. Lot No. 31057 1988. –/80 1TD 1W. 39.0 t.
DMS. Dia. DP249. Lot No. 31058 1988. –/80. 38.7 t.

Note: These units are owned by West Yorkshire PTE, although managed by Porterbrook Leasing Company.

155 341	**WY**	P	*NS*	NL	52341	57341
155 342	**WY**	P	*NS*	NL	52342	57342
155 343	**WY**	P	*NS*	NL	52343	57343
155 344	**WY**	P	*NS*	NL	52344	57344
155 345	**WY**	P	*NS*	NL	52345	57345
155 346	**WY**	P	*NS*	NL	52346	57346
155 347	**WY**	P	*NS*	NL	52347	57347

CLASS 156   SUPER SPRINTER   METRO-CAMMELL

DMSL–DMS.

Construction: Steel.
Engines: One Cummins NT855R5 of 213 kW (285 h.p.) at 2100 r.p.m. per car
Bogies: One P3-10 (powered) and one BT38 (non-powered).
Couplers: BSI.
Seating Layout: 2+2 facing/unidirectional.
Dimensions: 23.03 x 2.73 m.
Gangways: Throughout.   **Wheel Arrangement:** 2-B – B-2.
Doors: Single-leaf sliding plug.   **Maximum Speed:** 75 m.p.h.
Multiple Working: Classes 142, 143, 144, 150, 153, 155, 156, 158, 159, 170.

DMSL. Dia. DP244. Lot No. 31028 1988–89. –/74 (†* –/72, st –/70, u –/68) 1TD
1W. 36.1 t.
DMS. Dia. DP245. Lot No. 31029 1987–89. –/76 (q –/78, † –/74, tu§ –/72) 35.5 t.

Notes: 156 500–514 are owned by Strathclyde PTE, although managed by
Angel Train Contracts.
Units in **RE**, **RN** or **NS** livery have been fitted with new seating.

156 401	r*	**RE**	P	*CT*	TS	52401	57401
156 402	r*	**RE**	P	*CT*	TS	52402	57402
156 403	r*	**RE**	P	*CT*	TS	52403	57403
156 404	r*	**RE**	P	*CT*	TS	52404	57404
156 405	r*	**RE**	P	*CT*	TS	52405	57405
156 406	r*	**RE**	P	*CT*	TS	52406	57406
156 407	r*	**CT**	P	*CT*	TS	52407	57407
156 408	r*	**RE**	P	*CT*	TS	52408	57408
156 409	r*	**RE**	P	*CT*	TS	52409	57409
156 410	r*	**RE**	P	*CT*	TS	52410	57410
156 411	r*	**RE**	P	*CT*	TS	52411	57411
156 412	r*	**RE**	P	*CT*	TS	52412	57412
156 413	r*	**RE**	P	*CT*	TS	52413	57413
156 414	r*	**RE**	P	*CT*	TS	52414	57414
156 415	r*	**RE**	P	*CT*	TS	52415	57415
156 416	r*	**RE**	P	*CT*	TS	52416	57416
156 417	r*	**RE**	P	*CT*	TS	52417	57417
156 418	r*	**RE**	P	*CT*	TS	52418	57418
156 419	r*	**RE**	P	*CT*	TS	52419	57419
156 420	s	**RN**	P	*NW*	NH	52420	57420
156 421	s	**RN**	P	*NW*	NH	52421	57421
156 422	r*	**RE**	P	*CT*	TS	52422	57422
156 423	s	**RN**	P	*NW*	NH	52423	57423
156 424	s	**RN**	P	*NW*	NH	52424	57424
156 425	s	**RN**	P	*NW*	NH	52425	57425
156 426	s	**RN**	P	*NW*	NH	52426	57426
156 427	s	**RN**	P	*NW*	NH	52427	57427
156 428	s	**RN**	P	*NW*	NH	52428	57428
156 429	s	**RN**	P	*NW*	NH	52429	57429
156 430	t	**SC**	A	*SR*	CK	52430	57430

156 431	t	**SC**	A	*SR*	CK	52431	57431	
156 432	t	**SC**	A	*SR*	CK	52432	57432	
156 433	t	**SC**	A	*SR*	CK	52433	57433	The Kilmarnock Edition
156 434	t	**SC**	A	*SR*	CK	52434	57434	
156 435	t	**SC**	A	*SR*	CK	52435	57435	
156 436	†	**SC**	A	*SR*	CK	52436	57436	
156 437	rt	**SC**	A	*SR*	CK	52437	57437	
156 438		**PS**	A	*NS*	NL	52438	57438	
156 439	rt	**SC**	A	*SR*	CK	52439	57439	
156 440	s	**RN**	P	*NW*	NH	52440	57440	
156 441	s	**RN**	P	*NW*	NH	52441	57441	
156 442	rt	**SC**	A	*SR*	CK	52442	57442	
156 443	q	**NS**	A	*NS*	HT	52443	57443	
156 444	q	**NS**	A	*NS*	HT	52444	57444	
156 445	u	**SC**	A	*SR*	CK	52445	57445	
156 446	rt	**SR**	A	*SR*	CK	52446	57446	
156 447	ru	**SR**	A	*SR*	CK	52447	57447	
156 448	q	**NS**	A	*NS*	HT	52448	57448	
156 449	ru	**SR**	A	*SR*	CK	52449	57449	
156 450	t	**PS**	A	*SR*	CK	52450	57450	
156 451		**PS**	A	*NS*	HT	52451	57451	
156 452	s	**RN**	P	*NW*	NH	52452	57452	
156 453	ru	**SR**	A	*SR*	CK	52453	57453	
156 454	q	**NS**	A	*NS*	HT	52454	57454	
156 455	s	**RN**	P	*NW*	NH	52455	57455	
156 456	rt	**SR**	A	*SR*	CK	52456	57456	
156 457	rt	**SR**	A	*SR*	CK	52457	57457	
156 458	t	**SR**	A	*SR*	CK	52458	57458	
156 459	s	**RN**	P	*NW*	NH	52459	57459	
156 460	s	**RN**	P	*NW*	NH	52460	57460	
156 461	s	**RN**	P	*NW*	NH	52461	57461	
156 462	r	**PS**	A	*SR*	CK	52462	57462	
156 463	q	**NS**	A	*NS*	HT	52463	57463	
156 464	s	**RN**	P	*NW*	NH	52464	57464	
156 465	u	**PS**	A	*SR*	CK	52465	57465	Bonny Prince Charlie
156 466	s	**RN**	P	*NW*	NH	52466	57466	
156 467	r	**PS**	A	*SR*	CK	52467	57467	
156 468		**PS**	A	*NS*	NL	52468	57468	
156 469	q	**NS**	A	*NS*	HT	52469	57469	
156 470	q	**NS**	A	*NS*	NL	52470	57470	
156 471	q	**NS**	A	*NS*	NL	52471	57471	
156 472	q	**NS**	A	*NS*	NL	52472	57472	
156 473		**PS**	A	*NS*	NL	52473	57473	
156 474	rt	**SR**	A	*SR*	CK	52474	57474	
156 475	q	**NS**	A	*NS*	NL	52475	57475	
156 476	rt	**PS**	A	*SR*	CK	52476	57476	
156 477	rt	**SR**	A	*SR*	CK	52477	57477	
156 478	t	**SR**	A	*SR*	CK	52478	57478	
156 479	q	**NS**	A	*NS*	NL	52479	57479	
156 480	q	**NS**	A	*NS*	NL	52480	57480	
156 481	q	**NS**	A	*NS*	NL	52481	57481	

156 482	q	**NS**	A	*NS*	NL	52482	57482
156 483		**PS**	A	*NS*	NL	52483	57483
156 484	q	**NS**	A	*NS*	NL	52484	57484
156 485	ru	**PS**	A	*SR*	CK	52485	57485
156 486	q	**NS**	A	*NS*	NL	52486	57486
156 487	q	**NS**	A	*NS*	NL	52487	57487
156 488		**PS**	A	*NS*	NL	52488	57488
156 489	q	**NS**	A	*NS*	NL	52489	57489
156 490	q	**NS**	A	*NS*	NL	52490	57490
156 491	q	**NS**	A	*NS*	NL	52491	57491
156 492	r†	**SR**	A	*SR*	CK	52492	57492
156 493	rt	**SR**	A	*SR*	CK	52493	57493
156 494	§	**SC**	A	*SR*	CK	52494	57494
156 495	ru	**SC**	A	*SR*	CK	52495	57495
156 496	ru	**SR**	A	*SR*	CK	52496	57496
156 497		**PS**	A	*NS*	NL	52497	57497
156 498	q	**NS**	A	*NS*	NL	52498	57498
156 499	rt	**PS**	A	*SR*	CK	52499	57499
156 500	u	**SC**	A	*SR*	CK	52500	57500
156 501		**SC**	A	*SR*	CK	52501	57501
156 502		**SC**	A	*SR*	CK	52502	57502
156 503		**SC**	A	*SR*	CK	52503	57503
156 504		**SC**	A	*SR*	CK	52504	57504
156 505		**SC**	A	*SR*	CK	52505	57505
156 506		**SC**	A	*SR*	CK	52506	57506
156 507		**SC**	A	*SR*	CK	52507	57507
156 508		**SC**	A	*SR*	CK	52508	57508
156 509		**SC**	A	*SR*	CK	52509	57509
156 510		**SC**	A	*SR*	CK	52510	57510
156 511		**SC**	A	*SR*	CK	52511	57511
156 512		**SC**	A	*SR*	CK	52512	57512
156 513		**SC**	A	*SR*	CK	52513	57513
156 514		**SC**	A	*SR*	CK	52514	57514

CLASS 158/0   BREL

DMSL (B)–DMSL (A) or DMCL–DMSL or DMCL–MSL–DMSL.

Construction: Welded aluminium.
Engines:
158 701–158 814: One Cummins NTA855R of 260 kW (350 h.p.) at 1900 r.p.m. per car.
158 863–158 872: One Cummins NTA855R of 300 kW (400 h.p.) at 2100 r.p.m. per car.
158 815–158 862: One Perkins 2006-TWH of 260 kW (350 h.p.) at 1900 r.p.m. per car.
Bogies: One BREL P4 (powered) and one BREL T4 (non-powered) per car.
Couplers: BSI.
Seating Layout: 2+2 facing/unidirectional in standard class and in ScotRail first class. 2+2 facing in Northern Spirit first class, 2+1 facing/unidirectional in Virgin Cross-Country first class.

Dimensions: 22.57 x 2.70 m.
Gangways: Throughout.
Doors: Twin-leaf swing plug.
Multiple Working: Classes 142, 143, 144, 150, 153, 155, 156, 158, 159, 170.

Wheel Arrangement: 2-B – B-2.
Maximum Speed: 90 m.p.h.

DMSL (B). Dia. DP252. Lot No. 31051 BREL Derby 1989–92. –/68 1TD 1W. (†–/66 1TD 1W). Public telephone and trolley space. 38.5 t.
MSL. Dia. DR207. Lot No. 31050 BREL Derby 1991. 37.1 t. –/70 2T. 37.1 t.
DMSL (A). Dia. DP251 Lot No. 31052 BREL Derby 1989–92. –/70 († –/68; § 32/32) 1T. 37.8 t.

The above details refer to the "as built" condition. The following DMSL(B) have now been converted to DMCL as follows:

52701–52744 (Scotrail/Northern Spirit). Dia. DP318. 15/51 1TD 1W (*15/53 1TD 1W).
52747–52751. (Virgin Cross-Country). Dia. DP323. 9/51 1TD 1W.
52757–52759. (First North-Western). Dia. DP333. 16/51 1TD 1W.
52760–779/781. (Northern Spirit 2-car Units). Dia. DP331. 16/48 1TD 1W.
52798–814 (Northern Spirit 3-car Units). Dia. DP332. 32/32 1TD 1W.

s Refurbished with new shape seat cushions. Fitted with table lamps in first class. Units in **CT** livery have also been fitted with new shape seat cusions.
† Fitted with new seating.
Non-standard Livery: 158 867 is in prototype Wales & West Passenger Trains livery of grey, orange, light blue and dark blue with orange doors.

158 701	*	**SR**	P	*SR*	HA	52701	57701
158 702	*	**SR**	P	*SR*	HA	52702	57702	BBC Scotland – 75 Year
158 703	*	**SR**	P	*SR*	HA	52703	57703
158 704	*	**SR**	P	*SR*	HA	52704	57704
158 705	*	**SR**	P	*SR*	HA	52705	57705
158 706	*	**SR**	P	*SR*	HA	52706	57706
158 707		**SR**	P	*SR*	HA	52707	57707	Far North Line 125th ANNIVERSARY
158 708		**SR**	P	*SR*	HA	52708	57708
158 709		**SR**	P	*SR*	HA	52709	57709
158 710	*	**SR**	P	*SR*	HA	52710	57710
158 711	*	**SR**	P	*SR*	HA	52711	57711
158 712	*	**SR**	P	*SR*	HA	52712	57712
158 713	*	**SR**	P	*SR*	HA	52713	57713
158 714	*	**SR**	P	*SR*	HA	52714	57714
158 715		**SR**	P	*SR*	HA	52715	57715	Haymarket
158 716	*	**SR**	P	*SR*	HA	52716	57716
158 717	*	**SR**	P	*SR*	HA	52717	57717
158 718	*	**SR**	P	*SR*	HA	52718	57718
158 719	*	**SR**	P	*SR*	HA	52719	57719
158 720		**SR**	P	*SR*	HA	52720	57720
158 721	*	**SR**	P	*SR*	HA	52721	57721
158 722	*	**SR**	P	*SR*	HA	52722	57722
158 723	*	**SR**	P	*SR*	HA	52723	57723
158 724		**SR**	P	*SR*	HA	52724	57724
158 725	*	**SR**	P	*SR*	HA	52725	57725

158 726	*	**SR**	P	*SR*	HA	52726	57726	
158 727	*	**SR**	P	*SR*	HA	52727	57727	
158 728		**SR**	P	*SR*	HA	52728	57728	
158 729		**SR**	P	*SR*	HA	52729	57729	
158 730	*	**SR**	P	*SR*	HA	52730	57730	
158 731	*	**SR**	P	*SR*	HA	52731	57731	
158 732	*	**SR**	P	*SR*	HA	52732	57732	
158 733	*	**SR**	P	*SR*	HA	52733	57733	
158 734	*	**SR**	P	*SR*	HA	52734	57734	
158 735		**SR**	P	*SR*	HA	52735	57735	
158 736	*	**SR**	P	*SR*	HA	52736	57736	
158 737		**TX**	P	*NS*	NL	52737	57737	
158 738		**SR**	P	*SR*	HA	52738	57738	
158 739		**SR**	P	*SR*	HA	52739	57739	
158 740	*	**SR**	P	*SR*	HA	52740	57740	
158 741	*	**SR**	P	*SR*	HA	52741	57741	
158 742		**RE**	P	*NS*	NL	52742	57742	
158 743		**RE**	P	*NS*	NL	52743	57743	
158 744		**RE**	P	*NS*	NL	52744	57744	
158 745	†	**WW**	P	*WW*	CF	52745	57745	Pont Britannia
158 746	†	**WW**	P	*WW*	CF	52746	57746	
158 747		**RE**	P	*VX*	NH	52747	57747	
158 748		**RE**	P	*VX*	NH	52748	57748	
158 749		**RE**	P	*VX*	NH	52749	57749	
158 750		**RE**	P	*VX*	NH	52750	57750	
158 751		**RE**	P	*VX*	NH	52751	57751	
158 752		**NW**	P	*NW*	NH	52752	57752	
158 753		**NW**	P	*NW*	NH	52753	57753	
158 754		**NW**	P	*NW*	NH	52754	57754	
158 755		**NW**	P	*NW*	NH	52755	57755	
158 756		**NW**	P	*NW*	NH	52756	57756	
158 757		**NW**	P	*NW*	NH	52757	57757	
158 758		**NW**	P	*NW*	NH	52758	57758	
158 759		**NW**	P	*NW*	NH	52759	57759	
158 760	s	**TX**	P	*NS*	NL	52760	57760	
158 761	s	**TX**	P	*NS*	NL	52761	57761	
158 762	s	**TX**	P	*NS*	NL	52762	57762	
158 763	s	**TX**	P	*NS*	NL	52763	57763	
158 764	s	**TX**	P	*NS*	NL	52764	57764	
158 765	s	**TX**	P	*NS*	NL	52765	57765	
158 766	s	**TX**	P	*NS*	NL	52766	57766	
158 767	s	**TX**	P	*NS*	NL	52767	57767	
158 768	s	**TX**	P	*NS*	NL	52768	57768	
158 769	s	**TX**	P	*NS*	NL	52769	57769	
158 770	s	**TX**	P	*NS*	NL	52770	57770	
158 771	s	**TX**	P	*NS*	HT	52771	57771	
158 772	s	**TX**	P	*NS*	NL	52772	57772	
158 773	s	**TX**	P	*NS*	NL	52773	57773	
158 774	s	**TX**	P	*NS*	NL	52774	57774	
158 775	s	**TX**	P	*NS*	HT	52775	57775	
158 776	s	**TX**	P	*NS*	HT	52776	57776	

158 777	s	**TX**	P	*NS*	HT	52777		57777
158 778	s	**TX**	P	*NS*	HT	52778		57778
158 779	s	**TX**	P	*NS*	HT	52779		57779
158 781	s	**TX**	P	*NS*	HT	52781		57781
158 782	r	**RE**	A	*CT*	TS	52782		57782
158 783	r	**CT**	A	*CT*	TS	52783		57783
158 787	r	**CT**	A	*CT*	TS	52787		57787
158 788	r	**RE**	A	*CT*	TS	52788		57788
158 795	r	**CT**	A	*CT*	TS	52795		57795
158 798	s	**TX**	P	*NS*	HT	52798	58715	57798
158 799	s	**TX**	P	*NS*	HT	52799	58716	57799
158 800	s	**TX**	P	*NS*	HT	52800	58717	57800
158 801	s	**TX**	P	*NS*	HT	52801	58701	57801
158 802	s	**TX**	P	*NS*	HT	52802	58702	57802
158 803	s	**TX**	P	*NS*	HT	52803	58703	57803
158 804	s	**TX**	P	*NS*	HT	52804	58704	57804
158 805	s	**TX**	P	*NS*	HT	52805	58705	57805
158 806	s	**TX**	P	*NS*	HT	52806	58706	57806
158 807	s	**TX**	P	*NS*	HT	52807	58707	57807
158 808	s	**TX**	P	*NS*	HT	52808	58708	57808
158 809	s	**TX**	P	*NS*	HT	52809	58709	57809
158 810	s	**TX**	P	*NS*	HT	52810	58710	57810
158 811	s	**TX**	P	*NS*	HT	52811	58711	57811
158 812	s	**TX**	P	*NS*	HT	52812	58712	57812
158 813	s	**TX**	P	*NS*	HT	52813	58713	57813
158 814	s	**TX**	P	*NS*	HT	52814	58714	57814
158 815	†	**RE**	A	*WW*	CF	52815		57815
158 816	†	**RE**	A	*WW*	CF	52816		57816
158 817	†	**RE**	A	*WW*	CF	52817		57817
158 818	†	**RE**	A	*WW*	CF	52818		57818
158 819	†	**RE**	A	*WW*	CF	52819		57819
158 820	†	**RE**	A	*WW*	CF	52820		57820
158 821	†	**RE**	A	*WW*	CF	52821		57821
158 822	†	**RE**	A	*WW*	CF	52822		57822
158 823	†	**RE**	A	*WW*	CF	52823		57823
158 824	†	**RE**	A	*WW*	CF	52824		57824
158 825	†	**RE**	A	*WW*	CF	52825		57825
158 826	†	**RE**	A	*WW*	CF	52826		57826
158 827	†	**RE**	A	*WW*	CF	52827		57827
158 828	†	**RE**	A	*WW*	CF	52828		57828
158 829	†	**RE**	A	*WW*	CF	52829		57829
158 830	†	**RE**	A	*WW*	CF	52830		57830
158 831	†	**RE**	A	*WW*	CF	52831		57831
158 832	†	**RE**	A	*WW*	CF	52832		57832
158 833	†	**RE**	A	*WW*	CF	52833		57833
158 834	†	**RE**	A	*WW*	CF	52834		57834
158 835	†	**RE**	A	*WW*	CF	52835		57835
158 836	†	**RE**	A	*WW*	CF	52836		57836
158 837	†	**RE**	A	*WW*	CF	52837		57837
158 838	†	**RE**	A	*WW*	CF	52838		57838
158 839	†	**RE**	A	*WW*	CF	52839		57839

158 840	†	**RE**	A	*WW*	CF	52840 57840
158 841	†	**RE**	A	*WW*	CF	52841 57841
158 842	†	**RE**	A	*WW*	CF	52842 57842
158 843	†	**RE**	A	*WW*	CF	52843 57843
158 844	r	**CT**	A	*CT*	TS	52844 57844
158 845	r	**CT**	A	*CT*	TS	52845 57845
158 846	r	**CT**	A	*CT*	TS	52846 57846
158 847	r	**RE**	A	*CT*	TS	52847 57847
158 848	r	**CT**	A	*CT*	TS	52848 57848
158 849	r	**CT**	A	*CT*	TS	52849 57849
158 850	r	**CT**	A	*CT*	TS	52850 57850
158 851	r	**CT**	A	*CT*	TS	52851 57851
158 852	r	**CT**	A	*CT*	TS	52852 57852
158 853	r	**CT**	A	*CT*	TS	52853 57853
158 854	r	**RE**	A	*CT*	TS	52854 57854
158 855	r	**RE**	A	*CT*	TS	52855 57855
158 856	r	**RE**	A	*CT*	TS	52856 57856
158 857	r	**RE**	A	*CT*	TS	52857 57857
158 858	r	**RE**	A	*CT*	TS	52858 57858
158 859	r	**CT**	A	*CT*	TS	52859 57859
158 860	r	**CT**	A	*CT*	TS	52860 57860
158 861	r	**RE**	A	*CT*	TS	52861 57861
158 862	r	**RE**	A	*CT*	TS	52862 57862
158 863	†	**RE**	A	*WW*	CF	52863 57863
158 864	†	**RE**	A	*WW*	CF	52864 57864
158 865	†	**RE**	A	*WW*	CF	52865 57865
158 866	†	**RE**	A	*WW*	CF	52866 57866
158 867	†	**O**	A	*WW*	CF	52867 57867
158 868	†	**RE**	A	*WW*	CF	52868 57868
158 869	†	**RE**	A	*WW*	CF	52869 57869
158 870	†	**RE**	A	*WW*	CF	52870 57870
158 871	†	**RE**	A	*WW*	CF	52871 57871
158 872	†	**RE**	A	*WW*	CF	52872 57872

CLASS 158/9   BREL

DMSL–DMS. Units leased by West Yorkshire PTE. Details as for Class 158/0 except for seating layout and toilets.

DMSL.. Dia. DP252. Lot No. 31051 BREL Derby 1990–92. –/70 1TD 1W. Public telephone and trolley space. 38.1 t.
DMS. Dia. DP251. Lot No. 31052 BREL Derby 1990–92. –/72. 37.8 t.

Note: These units are leased by West Yorkshire PTE and are managed by Porterbrook Leasing Company.

158 901	**WY**	P	*NS*	NL	52901 57901
158 902	**WY**	P	*NS*	NL	52902 57902
158 903	**WY**	P	*NS*	NL	52903 57903
158 904	**WY**	P	*NS*	NL	52904 57904
158 905	**WY**	P	*NS*	NL	52905 57905
158 906	**WY**	P	*NS*	NL	52906 57906

158 907	**WY**	P	*NS*	NL	52907 57907
158 908	**WY**	P	*NS*	NL	52908 57908
158 909	**YN**	P	*NS*	NL	52909 57909
158 910	**WY**	P	*NS*	NL	52910 57910

CLASS 158/0   BREL

DMSL(B)–DMSL(A)–DMSL(A) or DMSL(B)–DMSL(A)–DMSL(B). Three car units formed by Central Trains from former 2-car units. For details see above.

158 951	r	**CT**	A	*CT*	TS	52780 57780	57784
158 952	r	**CT**	A	*CT*	TS	52785 57785	57786
158 953	r	**CT**	A	*CT*	TS	52790 57790	57796
158 954	r	**CT**	A	*CT*	TS	52791 57791	57786
158 955	r	**CT**	A	*CT*	TS	52792 57792	57796
158 956	r	**CT**	A	*CT*	TS	52793 57793	57784
158 957	r	**CT**	A	*CT*	TS	52794 57794	52789
158 958	r	**CT**	A	*CT*	TS	52797 57797	57789

CLASS 159   BREL

DMCL–MSL–DMSL. Built as Class 158. Converted before entering passenger service to Class 159 by Rosyth Dockyard.

Construction: Welded aluminium.
Engines: One Cummins NTA855R of 300 kW (400 h.p.) at 2100 r.p.m. per car.
Bogies: One BREL P4 (powered) and one BREL T4 (non-powered) per car.
Couplers: BSI.
Seating Layout: 2+2 facing/unidirectional (standard class), 2+1 facing (first class).
Dimensions: 23.21 x 2.82 m.
Gangways: Throughout.   **Wheel Arrangement:** 2-B – B-2 – B-2.
Doors: Twin-leaf swing plug.   **Maximum Speed:** 90 m.p.h.
Multiple Working: Classes 142, 143, 144, 150, 153, 155, 156, 158, 159, 170.

DMCL. Dia. DP322. Lot No. 31051 BREL Derby 1992–93. 24/28 1TD 1W. 38.5 t.
MSL. Dia. DR209. Lot No. 31050 BREL Derby 1992–93. 38 t. –/72 2T.
DMSL. Dia. DP260. Lot No. 31052 BREL Derby 1992–93. –/72 1T and parcels area. 37.8 t.

159 001	**NT** P	*SW*	SA	52873 58718 57873	CITY OF EXETER
159 002	**NT** P	*SW*	SA	52874 58719 57874	CITY OF SALISBURY
159 003	**SW** P	*SW*	SA	52875 58720 57875	TEMPLECOMBE
159 004	**NT** P	*SW*	SA	52876 58721 57876	BASINGSTOKE AND DEANE
159 005	**SW** P	*SW*	SA	52877 58722 57877	
159 006	**SW** P	*SW*	SA	52878 58723 57878	
159 007	**SW** P	*SW*	SA	52879 58724 57879	
159 008	**SW** P	*SW*	SA	52880 58725 57880	
159 009	**NT** P	*SW*	SA	52881 58726 57881	
159 010	**SW** P	*SW*	SA	52882 58727 57882	
159 011	**SW** P	*SW*	SA	52883 58728 57883	
159 012	**NT** P	*SW*	SA	52884 58729 57884	
159 013	**SW** P	*SW*	SA	52885 58730 57885	

159 014	**SW** P	*SW*	SA	52886	58731	57886
159 015	**SW** P	*SW*	SA	52887	58732	57887
159 016	**SW** P	*SW*	SA	52888	58733	57888
159 017	**NT** P	*SW*	SA	52889	58734	57889
159 018	**NT** P	*SW*	SA	52890	58735	57890
159 019	**NT** P	*SW*	SA	52891	58736	57891
159 020	**NT** P	*SW*	SA	52892	58737	57892
159 021	**NT** P	*SW*	SA	52893	58738	57893
159 022	**NT** P	*SW*	SA	52894	58739	57894

CLASS 165/0 NETWORK TURBO BREL

DMCL–DMS or DMCL–MS–DMS.

Construction: Welded aluminium.
Engines: One Perkins 2006-TWH of 260 kW (350 h.p.) at 1900 r.p.m. per car.
Bogies: BREL P3-17 (powered), BREL T3-17 (non-powered).
Couplers: BSI.
Seating Layout: 3+2 facing/unidirectional (standard class), 2+2 facing (first class).
Dimensions: 23.50 x 2.85 m.
Gangways: Within unit only.   **Wheel Arrangement:** 2-B (– B-2) – B-2.
Doors: Twin-leaf swing plug.   **Maximum Speed:** 75 m.p.h.
Multiple Working: Classes 165, 166, 168.

58801–58822. 58873–58878. DMCL. Dia. DP319. Lot No. 31087 BREL York 1990. 16/72 1T. 37.0 t.
58823–58833. DMCL. Dia. DP320. Lot No. 31089 BREL York 1991–92. 24/60 1T. 37.0 t.
MS. Dia. DR208. Lot No. 31090 BREL York 1991–92. –/106. 37.0 t.
DMS. Dia. DP253. Lot No. 31088 BREL York 1991–92. –/98. 37.0 t.

Note: 165 006–039 are fitted with tripcocks for working over London Underground tracks between Harrow-on-the-Hill and Amersham.

165 001	**NT**	A	*TT*	RG	58801	58834
165 002	**NT**	A	*TT*	RG	58802	58835
165 003	**NT**	A	*TT*	RG	58803	58836
165 004	**NT**	A	*TT*	RG	58804	58837
165 005	**NT**	A	*TT*	RG	58805	58838
165 006	**NT**	A	*CR*	AL	58806	58839
165 007	**NT**	A	*CR*	AL	58807	58840
165 008	**NT**	A	*CR*	AL	58808	58841
165 009	**NT**	A	*CR*	AL	58809	58842
165 010	**NT**	A	*CR*	AL	58810	58843
165 011	**NT**	A	*CR*	AL	58811	58844
165 012	**NT**	A	*CR*	AL	58812	58845
165 013	**NT**	A	*CR*	AL	58813	58846
165 014	**NT**	A	*CR*	AL	58814	58847
165 015	**NT**	A	*CR*	AL	58815	58848
165 016	**NT**	A	*CR*	AL	58816	58849
165 017	**NT**	A	*CR*	AL	58817	58850
165 018	**NT**	A	*CR*	AL	58818	58851

165 019	**NT**	A	*CR*	AL	58819		58852
165 020	**NT**	A	*CR*	AL	58820		58853
165 021	**NT**	A	*CR*	AL	58821		58854
165 022	**NT**	A	*CR*	AL	58822		58855
165 023	**NT**	A	*CR*	AL	58873		58867
165 024	**NT**	A	*CR*	AL	58874		58868
165 025	**NT**	A	*CR*	AL	58875		58869
165 026	**NT**	A	*CR*	AL	58876		58870
165 027	**NT**	A	*CR*	AL	58877		58871
165 028	**NT**	A	*CR*	AL	58878		58872
165 029	**NT**	A	*CR*	AL	58823	55404	58856
165 030	**NT**	A	*CR*	AL	58824	55405	58857
165 031	**NT**	A	*CR*	AL	58825	55406	58858
165 032	**NT**	A	*CR*	AL	58826	55407	58859
165 033	**NT**	A	*CR*	AL	58827	55408	58860
165 034	**NT**	A	*CR*	AL	58828	55409	58861
165 035	**NT**	A	*CR*	AL	58829	55410	58862
165 036	**NT**	A	*CR*	AL	58830	55411	58863
165 037	**NT**	A	*CR*	AL	58831	55412	58864
165 038	**NT**	A	*CR*	AL	58832	55413	58865
165 039	**NT**	A	*CR*	AL	58833	55414	58866

CLASS 165/1   NETWORK TURBO   BREL

DMCL–DMS or DMCL–MS–DMS.

Construction: Welded aluminium.
Engines: One Perkins 2006-TWH of 260 kW (350 h.p.) at 1900 r.p.m. per car.
Bogies: BREL P3-17 (powered), BREL T3-17 (non-powered).
Couplers: BSI.
Seating layout: 3+2 facing/unidirectional (standard class), 2+2 facing (first class).
Dimensions: 23.50 x 2.85 m.
Gangways: Within unit only.   **Wheel Arrangement:** 2-B (– B-2) – B-2.
Doors: Twin-leaf swing plug.   **Maximum Speed:** 90 m.p.h.
Multiple Working: Classes 165, 166, 168.

58953–58969. DMCL. Dia. DP320. Lot No. 31098 BREL York 1992. 16/66 1T. 37.0 t.
58879–58898. DMCL. Dia. DP319. Lot No. 31096 BREL York 1992. 16/72 1T. 37.0 t.
MS. Dia. DR208. Lot No. 31099 BREL 1992. –/106. 37.0 t.
DMS. Dia. DP253. Lot No. 31097 BREL 1992. –/98. 37.0 t.

165 101	**NT**	A	*TT*	RG	58953	55415	58916
165 102	**NT**	A	*TT*	RG	58954	55416	58917
165 103	**NT**	A	*TT*	RG	58955	55417	58918
165 104	**NT**	A	*TT*	RG	58956	55418	58919
165 105	**NT**	A	*TT*	RG	58957	55419	58920
165 106	**NT**	A	*TT*	RG	58958	55420	58921
165 107	**NT**	A	*TT*	RG	58959	55421	58922
165 108	**NT**	A	*TT*	RG	58960	55422	58923

165 109	**NT**	A	*TT*	RG	58961	55423	58924
165 110	**NT**	A	*TT*	RG	58962	55424	58925
165 111	**NT**	A	*TT*	RG	58963	55425	58926
165 112	**NT**	A	*TT*	RG	58964	55426	58927
165 113	**NT**	A	*TT*	RG	58965	55427	58928
165 114	**NT**	A	*TT*	RG	58966	55428	58929
165 116	**NT**	A	*TT*	RG	58968	55430	58931
165 117	**NT**	A	*TT*	RG	58969	55431	58932
165 118	**NT**	A	*TT*	RG	58879		58933
165 119	**NT**	A	*TT*	RG	58880		58934
165 120	**NT**	A	*TT*	RG	58881		58935
165 121	**NT**	A	*TT*	RG	58882		58936
165 122	**NT**	A	*TT*	RG	58883		58937
165 123	**NT**	A	*TT*	RG	58884		58938
165 124	**NT**	A	*TT*	RG	58885		58939
165 125	**NT**	A	*TT*	RG	58886		58940
165 126	**NT**	A	*TT*	RG	58887		58941
165 127	**NT**	A	*TT*	RG	58888		58942
165 128	**NT**	A	*TT*	RG	58889		58943
165 129	**NT**	A	*TT*	RG	58890		58944
165 130	**NT**	A	*TT*	RG	58891		58945
165 131	**NT**	A	*TT*	RG	58892		58946
165 132	**NT**	A	*TT*	RG	58893		58947
165 133	**NT**	A	*TT*	RG	58894		58948
165 134	**NT**	A	*TT*	RG	58895		58949
165 135	**NT**	A	*TT*	RG	58896		58950
165 136	**NT**	A	*TT*	RG	58897		58951
165 137	**NT**	A	*TT*	RG	58898		58952
Spare	**NT**	A	*TT*	ZC (S)		55429	58930

CLASS 166   NETWORK EXPRESS TURBO   ABB

DMCL (A)–MS–DMCL (B). Built for Paddington–Oxford/Newbury services. Air conditioned.

Construction: Welded aluminium.
Engines: One Perkins 2006-TWH of 260 kW (350 h.p.) at 1900 r.p.m. per car.
Bogies: BREL P3-17 (powered), BREL T3-17 (non-powered).
Couplers: BSI.
Seating Layout: 3+2 facing/unidirectional (standard class) with 20 standard class seats in 2+2 format in DMCL(B), 2+2 facing (first class).
Dimensions: 23.50 x 2.85 m.
Gangways: Within unit only.   **Wheel Arrangement:** 2-B – B-2 – B-2.
Doors: Twin-leaf swing plug.   **Maximum Speed:** 90 m.p.h.
Multiple Working: Classes 165, 166, 168.

DMCL (A). Dia. DP321. Lot No. 31116 ABB York 1992–3. 16/75 1T. 40.62 t.
MS. Dia. DR209. Lot No. 31117 ABB York 1992–93. –/96. 38.04 t.
DMCL (B). Dia. DP321. Lot No. 31116 ABB York 1992–93. 16/72 1T. 40.64 t.

166 201	**TT**	A	*TT*	RG	58101	58601	58122
166 202	**TT**	A	*TT*	RG	58102	58602	58123
166 203	**NT**	A	*TT*	RG	58103	58603	58124
166 204	**NT**	A	*TT*	RG	58104	58604	58125
166 205	**NT**	A	*TT*	RG	58105	58605	58126
166 206	**TT**	A	*TT*	RG	58106	58606	58127
166 207	**NT**	A	*TT*	RG	58107	58607	58128
166 208	**NT**	A	*TT*	RG	58108	58608	58129
166 209	**NT**	A	*TT*	RG	58109	58609	58130
166 210	**NT**	A	*TT*	RG	58110	58610	58131
166 211	**NT**	A	*TT*	RG	58111	58611	58132
166 212	**NT**	A	*TT*	RG	58112	58612	58133
166 213	**NT**	A	*TT*	RG	58113	58613	58134
166 214	**NT**	A	*TT*	RG	58114	58614	58135
166 215	**NT**	A	*TT*	RG	58115	58615	58136
166 216	**NT**	A	*TT*	RG	58116	58616	58137
166 217	**NT**	A	*TT*	RG	58117	58617	58138
166 218	**NT**	A	*TT*	RG	58118	58618	58139
166 219	**NT**	A	*TT*	RG	58119	58619	58140
166 220	**NT**	A	*TT*	RG	58120	58620	58141
166 221	**NT**	A	*TT*	RG	58121	58621	58142

CLASS 168   CLUBMAN   ADTRANZ

DMSL (A)–MSL–MS–DMSL (B). Air conditioned.

Construction: Welded aluminium bodies with bolt-on steel ends.
Engines: One MTU 6R183TD13H of 315 kW (422 h.p.) at 1900 r.p.m. per car.
Transmission: Hydraulic. Voith T211rzze to ZF final drive.
Bogies: One Adtranz P3–23 and one BREL T3–23 per car.
Couplers: BSI.
Seating Layout: 2+2 facing/unidirectional.
Dimensions: 23.62 x 2.69 m (driving cars), 23.61 x 2.69 m (centre cars).
Gangways: Within unit only.   **Wheel Arrangement:** 2-B (– B-2 – B-2) – B-2.
Doors: Twin-leaf swing plug.   **Maximum Speed:** 100 m.p.h.
Multiple Working: Classes 165, 166, 168.

58151–58155. DMSL(A). Dia. DP270. Adtranz Derby 1997–98. –/60 1TD 1W. 43.7 t.
58156–58163. DMSL(A). Dia. DP280. Adtranz Derby 2000. –/59 1TD 2W. 43.7 t.
58651–58655. MSL. Dia. DR211. Adtranz Derby 1998. –/73 1T. 41.0 t.
58656–58660. MS. Dia. DR211. Adtranz Derby 1998. –/77. 40.5 t.
58661–58663. MS. Dia. DR211. Adtranz Derby 2000. –/76. 42.4 t.
58251–58255. DMSL(B). Dia. DP270. Adtranz Derby 1998. –/66 1T. 43.6 t.
58256–58263. DMSL(B). Dia. DP281. Adtranz Derby 2000. –/69 1T. 43.6 t.

Notes:

Fitted with tripcocks for working over London Underground tracks between Harrow-on-the-Hill and Amersham.
58656–60 were formerly numbered 58451–6.

168 001	**CR**	P	*CR*	AL	58151	58651	58251
168 002	**CR**	P	*CR*	AL	58152	58652	58252
168 003	**CR**	P	*CR*	AL	58153	58653	58253
168 004	**CR**	P	*CR*	AL	58154	58654	58254
168 005	**CR**	P	*CR*	AL	58155	58655	58255
168 106	**CR**	P	*CR*	AL	58156	58656	58256
168 107	**CR**	P	*CR*	AL	58157	58657	58257
168 108	**CR**	P	*CR*	AL	58158	58658	58258
168 109	**CR**	P	*CR*	AL	58159	58659	58259
168 110	**CR**	P	*CR*	AL	58160	58660	58260
168 111	**CR**	H	*CR*	AL	58161	58661	58261
168 112	**CR**	H	*CR*	AL	58162	58662	58262
168 113	**CR**	H	*CR*	AL	58163	58663	58263

CLASS 170   TURBOSTAR   ADTRANZ

Various formations. Air conditioned.

Construction: Welded aluminium bodies with bolt-on steel ends.
Engines: One MTU 6R183TD13H of 315 kW (422 h.p.) at 1900 r.p.m. per car.
Transmission: Hydraulic. Voith T211rzze to ZF final drive.
Bogies: One Adtranz P3–23 and one BREL T3–23 per car.
Couplers: BSI.
Seating Layout: facing/unidirectional (2+2 in standard class and in first class on Class 170/1, 2+1 in first class on Class 170/2 and 170/4).
Dimensions: 23.62 x 2.69 m (driving cars), 23.61 x 2.69 m (centre cars).
Gangways: Within unit only.   **Wheel Arrangement:** 2-B (– B-2) – B-2.
Doors: Twin-leaf swing plug.   **Maximum Speed:** 100 m.p.h.
Multiple Working: Classes 142, 143, 144, 150, 153, 155, 156, 158, 159, 170.

Class 170/1. Midland Mainline Units. DMCL–DMCL.

DMCL (A). Dia. DP324. Adtranz Derby 1998–1999. 12/45 1TD 2W. 45.19 t.
DMCL (B). Dia. DP325. Adtranz Derby 1998–1999. 12/52 1T. Catering point. 45.22 t

170 101	**MM**	P	*MM*	DY	50101	79101
170 102	**MM**	P	*MM*	DY	50102	79102
170 103	**MM**	P	*MM*	DY	50103	79103
170 104	**MM**	P	*MM*	DY	50104	79104
170 105	**MM**	P	*MM*	DY	50105	79105
170 106	**MM**	P	*MM*	DY	50106	79106
170 107	**MM**	P	*MM*	DY	50107	79107
170 108	**MM**	P	*MM*	DY	50108	79108
170 109	**MM**	P	*MM*	DY	50109	79109
170 110	**MM**	P	*MM*	DY	50110	79110
170 111	**MM**	P	*MM*	DY	50111	79111
170 112	**MM**	P	*MM*	DY	50112	79112
170 113	**MM**	P	*MM*	DY	50113	79113
170 114	**MM**	P	*MM*	DY	50114	79114
170 115	**MM**	P	*MM*	DY	50115	79115
170 116	**MM**	P	*MM*	DY	50116	79116
170 117	**MM**	P	*MM*	DY	50117	79117

MCLRMB. Dia. DR3 . Adtranz Derby 2001. xx/xx 1T + carvery bar. . t.
These cars will be formed into ten Midland mainline sets in 2001.

| 55101 | 55103 | 55105 | 55107 | 55109 |
| 55102 | 55104 | 55106 | 55108 | 55110 |

Class 170/2. Anglia Railways Units. DMCL–MSLRB–DMSL.

DMCL. Dia. DP326. Adtranz Derby 1999. 30/3 1TD 2W. 44.30 t.
MSLRB. Dia. DR212. Adtranz Derby 1999. –/58 1T. Buffet and conductor's office 42.76 t.
DMSL. Dia. DP274. Adtranz Derby 1999. –/66 1T. 44.70 t.

170 201	**AN**	P	*AR*	NC	50201	56201	79201
170 202	**AN**	P	*AR*	NC	50202	56202	79202
170 203	**AN**	P	*AR*	NC	50203	56203	79203
170 204	**AN**	P	*AR*	NC	50204	56204	79204
170 205	**AN**	P	*AR*	NC	50205	56205	79205
170 206	**AN**	P	*AR*	NC	50206	56206	79206
170 207	**AN**	P	*AR*	NC	50207	56207	79207
170 208	**AN**	P	*AR*	NC	50208	56208	79208

Class 170/3. South West Trains Units. DMCL–DMCL.

DMCL(A). Dia. DP329. Adtranz Derby 2000. 9/43 1TD 2W. 45.80 t.
DMCL(B). Dia. DP330. Adtranz Derby 2000. 9/53 1T. 45.80 t.

170 301	**SW**	P	*SW*	SA	50301	79301
170 302	**SW**	P	*SW*	SA	50302	79302
170 303	**SW**	P	*SW*	SA	50303	79303
170 304	**SW**	P	*SW*	SA	50304	79304
170 305	**SW**	P	*SW*	SA	50305	79305
170 306	**SW**	P	*SW*	SA	50306	79306
170 307		P			50307	79307
170 308		P			50308	79308

Class 170/4. ScotRail Units. DMCL–MS–DMCL. Under construction.

DMCL(A). Dia. DP329. Adtranz Derby 1999–2001. 9/43 1TD 2W. 45.80 t.
MS. Dia. DR213. Adtranz Derby 1999–2001. –/76. 43.00 t.
DMCL(B). Dia. DP330. Adtranz Derby 1999–2001. 9/53 1T. 45.80 t.

170 401	**SR**	P	*SR*	HA	50401	56401	79401
170 402	**SR**	P	*SR*	HA	50402	56402	79402
170 403	**SR**	P	*SR*	HA	50403	56403	79403
170 404	**SR**	P	*SR*	HA	50404	56404	79404
170 405	**SR**	P	*SR*	HA	50405	56405	79405
170 406	**SR**	P	*SR*	HA	50406	56406	79406
170 407	**SR**	P	*SR*	HA	50407	56407	79407
170 408	**SR**	P	*SR*	HA	50408	56408	79408
170 409	**SR**	P	*SR*	HA	50409	56409	79409
170 410	**SR**	P	*SR*	HA	50410	56410	79410
170 411	**SR**	P	*SR*	HA	50411	56411	79411
170 412	**SR**	P	*SR*	HA	50412	56412	79412
170 413	**SR**	P	*SR*	HA	50413	56413	79413
170 414	**0**	P	*SR*	HA	50414	56414	79414

170 415	**0**	P	*SR*	HA	50415	56415	79415
170 416	**SR**	H	*SR*	HA	50416	56416	79416
170 417	**SR**	H	*SR*	HA	50417	56417	79417
170 418	**SR**	H	*SR*	HA	50418	56418	79418
170 419	**SR**	H	*SR*	HA	50419	56419	79419
170 420	**SR**	H	*SR*	HA	50420	56420	79420
170 421	**SR**	H	*SR*	HA	50421	56421	79421
170 422	**SR**	H	*SR*	HA	50422	56422	79422
170 423	**SR**	H	*SR*	HA	50423	56423	79423
170 424	**SR**	H	*SR*	HA	50424	56424	79424

Class 170/47. Units leased by ScotRail for Strathclyde PTE services. DMSL(A)–MS–DMSL(B). Under construction.

DMSL(A). Dia. DP284. Adtranz Derby 2001. –/55 1TD 2W. 45.80 t.
MS. Dia. DR219. Adtranz Derby 2001. –/76. 43.00 t.
DMSL(B). Dia. DP285. Adtranz Derby 1999–2001. –/67 1T. 45.80 t.

| 170 470 | **SP** | P | | | 50470 | 56470 | 79470 |
| 170 471 | **SP** | P | | | 50471 | 56471 | 79471 |

Class 170/5. Central Trains 2-car Units. DMSL–DMSL.

DMSL(A). Dia. DP275. Adtranz Derby 1999–2000. –/55 1TD 2W. 45.80 t.
DMSL(B). Dia. DP276. Adtranz Derby 1999–2000. –/73 1T. 46.80 t.

170 501	r	**CT**	P	*CT*	TS	50501 79501
170 502	r	**CT**	P	*CT*	TS	50502 79502
170 503	r	**CT**	P	*CT*	TS	50503 79503
170 504	r	**CT**	P	*CT*	TS	50504 79504
170 505	r	**CT**	P	*CT*	TS	50505 79505
170 506	r	**CT**	P	*CT*	TS	50506 79506
170 507	r	**CT**	P	*CT*	TS	50507 79507
170 508	r	**CT**	P	*CT*	TS	50508 79508
170 509	r	**CT**	P	*CT*	TS	50509 79509
170 510	r	**CT**	P	*CT*	TS	50510 79510
170 511	r	**CT**	P	*CT*	TS	50511 79511
170 512	r	**CT**	P	*CT*	TS	50512 79512
170 513	r	**CT**	P	*CT*	TS	50513 79513
170 514	r	**CT**	P	*CT*	TS	50514 79514
170 515	r	**CT**	P	*CT*	TS	50515 79515
170 516	r	**CT**	P	*CT*	TS	50516 79516
170 517	r	**CT**	P	*CT*	TS	50517 79517
170 518	r	**CT**	P	*CT*	TS	50518 79518
170 519	r	**CT**	P	*CT*	TS	50519 79519
170 520	r	**CT**	P	*CT*	TS	50520 79520
170 521	r	**CT**	P	*CT*	TS	50521 79521
170 522	r	**CT**	P	*CT*	TS	50522 79522
170 523	r	**CT**	P	*CT*	TS	50523 79523

Class 170/6. Central Trains 3-car Units. DMSL–MS–DMSL.

DMSL(A). Dia. DP275. Adtranz Derby 2000. –/55 1TD 2W. 45.80 t.
MS. Dia. DR214. Adtranz Derby 2000. –/74. 43.00 t.
DMSL(B). Dia. DP276. Adtranz Derby 2000. –/73 1T. 46.80 t.

170 630	r	**CT**	P	*CT*	TS	50630	56630	79630
170 631	r	**CT**	P	*CT*	TS	50631	56631	79631
170 632	r	**CT**	P	*CT*	TS	50632	56632	79632
170 633	r	**CT**	P	*CT*	TS	50633	56633	79633
170 634	r	**CT**	P	*CT*	TS	50634	56634	79634
170 635	r	**CT**	P	*CT*	TS	50635	56635	79635
170 636	r	**CT**	P	*CT*	TS	50636	56636	79636
170 637	r	**CT**	P	*CT*	TS	50637	56637	79637
170 638	r	**CT**	P	*CT*	TS	50638	56638	79638
170 639	r	**CT**	P	*CT*	TS	50639	56639	79639

CLASS 175   CORADIA 1000   ALSTOM

Construction: Steel.
Engines: One Cummins N14 of 335 kW (450 h.p.).
Transmission: Hydraulic. Voith T211rzze to ZF final drive.
Bogies:
Couplers: Scharfenberg.
Seating Layout: 2+2 facing/unidirectional.
Dimensions: 23.71 x 2.73 m (driving cars), 23.03 x 2.73 m (centre cars).
Gangways: Within unit only.
Wheel Arrangement: 2-B (– B-2 – B-2) – B-2.
Doors: Single-leaf swing plug.
Maximum Speed: 100 m.p.h.
Multiple Working: Classes 175, 180.

Class 175/0. DMSL–DMSL. 2-car units.

DMSL(A). Dia. DP278. Alstom Birmingham 1999–2000. –/54 1TD 2W. 51.00 t.
DMSL(B). Dia. DP279. Alstom Birmingham 1999–2000. –/64 1T. 51.00 t.

175 001	**FN**	A	*NW*	CH	50701	79701
175 002	**FN**	A	*NW*	CH	50702	79702
175 003	**FN**	A	*NW*	CH	50703	79703
175 004	**FN**	A	*NW*	CH	50704	79704
175 005	**FN**	A	*NW*	CH	50705	79705
175 006	**FN**	A	*NW*	CH	50706	79706
175 007	**FN**	A	*NW*	CH	50707	79707
175 008	**FN**	A	*NW*	CH	50708	79708
175 009	**FN**	A	*NW*	CH	50709	79709
175 010	**FN**	A	*NW*	CH	50710	79710
175 011	**FN**	A	*NW*	CH	50711	79711

Class 175/1. DMSL(A)–MSL–DMSL(B). 3-car units.

DMSL(A). Dia. DP278. Alstom Birmingham 1999–2001. –/54 1TD 2W. 51.00 t.
MSL. Dia. DR216. Alstom Birmingham 1999–2001. –/68 1T. 43 t 68 1T. 47.50 t.
DMSL(B). Dia. DP279. Alstom Birmingham 1999–2001. –/64 1T. 51.00 t.

175 101	**FN**	A		KR	50751	56751	79751
175 102	**FN**	A	*NW*	CH	50752	56752	79752
175 103	**FN**	A	*NW*	CH	50753	56753	79753
175 104	**FN**	A	*NW*	CH	50754	56754	79754

175 105	**FN**	A	*NW*	CH	50755	56755	79755
175 106	**FN**	A	*NW*	CH	50756	56756	79756
175 107	**FN**	A			50757	56757	79757
175 108	**FN**	A			50758	56758	79758
175 109	**FN**	A			50759	56759	79759
175 110	**FN**	A			50760	56760	79760
175 111	**FN**	A			50761	56761	79761
175 112	**FN**	A			50762	56762	79762
175 113	**FN**	A			50763	56763	79763
175 114	**FN**	A			50764	56764	79764
175 115	**FN**	A			50765	56765	79765
175 116	**FN**	A			50766	56766	79766

CLASS 180   CORADIA 1000   ALSTOM

New units under construction for First Great Western.

Construction: Steel.
Engines: One Cummins QSK19 of 560 kW (750 h.p.) at 2100 r.p.m.
Transmission: Hydraulic. Voith T312br to Voith final drive.
Bogies: Alstom MB2.
Couplers: Scharfenberg.
Seating Layout: facing/unidirectional (2+2 in standard class and 2+1 in first class.
Dimensions: 23.71 x 2.73 m (driving cars), 23.03 x 2.73 m (centre cars).
Gangways: Within unit only.
Wheel Arrangement: 2-B (– B-2 – B-2 – B–2) – B-2.
Doors: Single-leaf swing plug.
Maximum Speed: 125 m.p.h.
Multiple Working: Classes 175, 180.

DMSL(A). Dia. DP2   . Alstom Birmingham 2000–01. –/46 2W 1TD. 53.00 t.
MFL. Dia. DR1   . Alstom Birmingham 2000–01. 42/– 1T 1W + catering point. 51.50 t.
MSL. Dia. DR2   . Alstom Birmingham 2000–01. –/68 1T. 51.50 t.
MSLRB. Dia. DR2   . Alstom Birmingham 2000–01. –/56 1T. 51.50 t.
DMSL(B). Dia. DP276. Alstom Birmingham 2000–01. –/56 1T. 53.00 t.

180 101	**FW**	W		OM	50901	54901	55901	56901	59901
180 102		W			50902	54902	55902	56902	59902
180 103		W			50903	54903	55903	56903	59903
180 104		W			50904	54904	55904	56904	59904
180 105		W			50905	54905	55905	56905	59905
180 106		W			50906	54906	55906	56906	59906
180 107		W			50907	54907	55907	56907	59907
180 108		W			50908	54908	55908	56908	59908
180 109		W			50909	54909	55909	56909	59909
180 110		W			50910	54910	55910	56910	59910
180 111		W			50911	54911	55911	56911	59911
180 112		W			50912	54912	55912	56912	59912
180 113		W			50913	54913	55913	56913	59913
180 114		W			50914	54914	55914	56914	59914

2. DIESEL ELECTRIC UNITS

The following features are standard to ex-BR Southern Region diesel-electric multiple unit power cars (Classes 201–207):

Construction: Steel.
Engine: One English Electric 4SRKT Mk. 2 of 450 kW (600 h.p.) at 850 r.p.m.
Main Generator: English Electric EE824.
Traction Motors: Two English Electric EE507 mounted on the inner bogie.
Bogies: SR Mk. 4. (Former EMU TSL vehicles have Commonwealth bogies).
Couplers: Drophead buckeye.
Doors: Manually operated slam.
Brakes: Electro-pneumatic and automatic air.
Maximum Speed: 75 m.p.h.
Multiple Working: Other ex BR Southern Region DEMU vehicles.

CLASS 201/202   PRESERVED 'HASTINGS' UNIT

DMBS–2TSL–TSRB–TSL–DMBS.

Preserved unit made up from 2 Class 201 short-frame cars and 2 Class 202 long-frame cars. The 'Hastings' units were made with narrow body-profiles for use on the section between Tonbridge and Battle which had tunnels of restricted loading gauge. These tunnels were converted to single track operation in the 1980s thus allowing standard loading gauge stock to be used. The set also contains two former EMU trailers (not Hastings line gauge).
Gangways: Within unit only.
Seating Layout: 2+2 facing.
Dimensions: 18.36 x 2.50 m (60000/60501), 20.34 x 2.50 m. (60118/60529) 20.34 x 2.82 m (69337/70262).

60000. DMBS. Dia DB203. Lot No. 30329 1957. –/22. 54 t.
60501. TSL. Dia DB204. Lot No. 30331 1957. –/52 2T. 29 t.
70262. TSL (ex Class 411/5 EMU). Dia. DH208. Lot No. 30455 1958–99. –/64 2T. 33.78 t.
69337. TSRB (ex Class 422 EMU). Dia. DH209. Lot No. 30805 York 1970. –/40. 35 t.
60529. TSL. Dia DH203. Lot No. 30397 1957. –/60 2T. 30 t.
60118. DMBS. Dia DB203. Lot No. 30395 1957. –/30. 55 t.

201 001   **G** HD *ON* SE   60000 60501 70262 69337 60529 60118

Names:

60000	Hastings	60118	Tunbridge Wells

CLASS 205/0 (3H)   'HAMPSHIRE'   BR EASTLEIGH

DMBS–TSL–DTCsoL or DMBS–DTCsoL.

Gangways: Non-gangwayed.
Seating Layout: 3+2 facing or compartments.
Dimensions: 20.33 x 2.82 m (DMBS), 20.28 x 2.82 m (TS), 20.36 x 2.82 m (DTCsoL).

60111/117/154. DMBS. Dia DB203. Lot No. 30332 1957. –/52. 56 t.
60122–124. DMBS. Dia DB203. Lot No. 30540 1958–59. –/52. 56 t.
60146–151. DMBS. Dia DB204. Lot No. 30671 1960–62. –/42. 56 t.
60650–670. TS. Dia DH203. Lot No. 30542 1958–59. –/104. 30 t.
60673–678. TS. Dia DH203. Lot No. 30672 1960–62. –/104. 30 t.
60800. DTCsoL. Dia DE301. Lot No. 30333 1956–57. 13/50 2T. 32 t.
60811. DTCsoL. Dia DE302. Lot No. 30333 1956–57. 19/50 2T. 32 t.
60820. DTCsoL. Dia DE301. Lot No. 30399 1957–58. 13/50 2T. 32 t.
60823/824. DTCsoL. Dia DE301. Lot No. 30541 1958–59. 13/50 2T. 32 t.
60827–832. DTCsoL. Dia DE303. Lot No. 30673 1960–62. 13/62 2T. 32 t.

205 001	**CX** P	*SC*	SU	60154		60800
205 009	**CX** P	*SC*	SU	60108	60658	60808
205 012	**CX** P	*SC*	SU	60111		60811
205 018	**CX** P	*SC*	SU	60117	60674	60828
205 024	**N** P	*SC*	SU (S)	60123		60823
205 025	**CX** P	*SC*	SU	60124		60824
205 028	**CX** P	*SC*	SU	60146	60673	60827
205 032	**CX** P	*SC*	SU	60150		60831
205 033	**CX** P	*SC*	SU	60151	60678	60832
Spare	**CX** P	*SC*	ZG (S)		60650	
Spare	**G** HD	*ON*	SE	60122	60668	
Spare	**N** P	*SC*	SU (S)		60670	
Spare	**N** P	*SC*	SU (S)		60677	

CLASS 205/2 (3H)   'HAMPSHIRE'   BR EASTLEIGH

DMBS–TSL (ex Class 411/5 EMU)–DTSL. Refurbished 1980. Fluorescent lighting. PA.

Details as for Class 205/0 except:

Gangways: Within unit only.
Seating Layout: 3+2 facing.

DMBS. Dia. DB203. Lot No. 30332 1957. –/39. 57 t.
TSL. Dia. DH207. Converted from loco-hauled TS 4059 Lot No. 30149 Swindon 1955–57. –/64 2T. 33.78 t.
DTSL. Dia. DE204. Lot No. 30333 1957. –/76 2T. 32 t.
Note: This unit operates as a two-car set in winter.

205 205	**CX** P	*SC*	SU	60110	71634	60810

CLASS 207/0 (2D)   'OXTED'   BR EASTLEIGH

DMBS–DTS (formerly DMBS–TCsoL–DTS).

This class was built for the Oxted line and therefore referred to as 'Oxted' units. They were made with a narrower body-profile which also allowed them to be used through the restricted loading-gauge Somerhill Tunnel between Tonbridge and Grove Junction (Tunbridge Wells). This tunnel was converted to single track operation in the 1980s thus allowing standard loading gauge stock to be used.

Gangways: Non-gangwayed.
Seating Layout: 3+2 facing or compartments.
Dimensions: 20.33 x 2.74 m. (DMBS/TCsoL), 20.32 x 2.74 m. (DTS).

DMBS. Dia DB205. Lot No. 30625 1962. –/42. 56 t.
60616. TCsoL. Dia DH301. Lot No. 30626 1962. 24/42 1T. 31 t.
60916. DTS. Dia DE201. Lot No. 30627 1962. –/76. 32 t.

207 017	**CX**	P	*SC*	SU	60142	60916
Spare	**G**	HD	*ON*	SE	60138 60616	

CLASS 207/1 (3D)   'OXTED'   BR EASTLEIGH

DMBS–TSL–DTS.
Gangwayed sets with a Class 411 EMU trailer in the centre.

Gangways: Within unit only.
Seating Layout: 2+2 facing.
Dimensions: 20.34 x 2.74 m. (DMBS), 20.32 x 2.74 m. (DTS).

DMBS. Dia DB205. Lot No. 30625 1962. –/40. 56 t.
70286. TSL. Dia. DH206. Lot No. 30455 1958–59. –/64 2T. 33.78 t.
70547/9. TSL. Dia. DH206. Lot No. 30620 1960–61 –/64 2T. 33.78 t.
DTS. Dia DE201. Lot No. 30627 1962. –/75. 32 t.
Note: These units operate as two-car sets in winter.

207 201	**CX**	P	*SC*	SU	60129	70286	60901
207 202	**CX**	P	*SC*	SU	60130	70549	60904
207 203	**CX**	P	*SC*	SU	60127	70547	60903

CLASS 220   VOYAGER   BOMBARDIER

DMS–MSRB–MS–DMF. New units under construction for Virgin Cross-Country.

Construction: Steel.
Engine: Cummins of 750 h.p. (560 kW) at 1800 r.p.m.
Transmission: Two Alstom Onix 800 three-phase traction motors of 275 kW.
Braking: Rheostatic and electro-pneumatic.
Bogies: Bombardier B5005.
Couplers: Dellner.
Seating Layout: 2+2 mainly unidirectional (standard class, 2+1 facing/
unidirectional (first class).
Dimensions: 23.85 x 2.73 m. (outer cars), 22.82 x 2.73 m. (inner cars).
Gangways: Within unit only.
Wheel Arrangement: 2-Bo – Bo-2 – Bo-2 – Bo-2.
Doors: Single-leaf swing plug.
Maximum Speed: 125 m.p.h.
Multiple Working: Classes 220, 221.

DMS. Dia DC201. Bombardier Prorail 2000–01. –/42 1TD 1W.   .   t.
MSRB. Dia. DD201. Bombardier Prorail 2000–01. –/58.   .   t.
MS. Dia. DD202. Bombardier Prorail 2000–01. –/62 1TD 1W.   .   t.
DMF. Dia DC101. Bombardier Prorail 2000–01. 26/– 1TD 1W.   .   t.

220 001	**VT**	HX			60301	60201	60701	60401

220 002	**VT**	HX		60302	60202	60702	60402
220 003	**VT**	HX		60303	60203	60703	60403
220 004	**VT**	HX		60304	60204	60704	60404
220 005	**VT**	HX		60305	60205	60705	60405
220 006		HX		60306	60206	60706	60406
220 007		HX		60307	60207	60707	60407
220 008		HX		60308	60208	60708	60408
220 009		HX		60309	60209	60709	60409
220 010		HX		60310	60210	60710	60410
220 011		HX		60311	60211	60711	60411
220 012		HX		60312	60212	60712	60412
220 013		HX		60313	60213	60713	60413
220 014		HX		60314	60214	60714	60414
220 015		HX		60315	60215	60715	60415
220 016		HX		60316	60216	60716	60416
220 017		HX		60317	60217	60717	60417
220 018		HX		60318	60218	60718	60418
220 019		HX		60319	60219	60719	60419
220 020		HX		60320	60220	60720	60420
220 021		HX		60321	60221	60721	60421
220 022		HX		60322	60222	60722	60422
220 023		HX		60223	60223	60723	60423
220 024		HX		60224	60224	60724	60424
220 025		HX		60225	60225	60725	60425
220 026		HX		60226	60226	60726	60426
220 027		HX		60227	60227	60727	60427
220 028		HX		60228	60228	60728	60428
220 029		HX		60229	60229	60729	60429
220 030		HX		60330	60230	60730	60430
220 031		HX		60331	60231	60731	60431
220 032		HX		60332	60232	60732	60432
220 033		HX		60333	60233	60733	60433
220 034		HX		60334	60234	60734	60434

CLASS 221   VOYAGER   BOMBARDIER

DMS–MSRB–MS(–MS)–DMF. New tilting units under construction for Virgin Cross-Country (five-car units) and Virgin West Coast (four-car units).

Construction: Steel.
Engine: Cummins of 750 h.p. (560 kW) at 1800 r.p.m.
Transmission: Two Alstom Onix 800 three-phase traction motors of 275 kW.
Braking: Rheostatic and electro-pneumatic.
Bogies: Bombardier HVP.
Couplers: Dellner.
Seating Layout: 2+2 mainly unidirectional (standard class, 2+1 facing/unidirectional (first class).
Dimensions: 23.85 x 2.73 m. (outer cars), 22.82 x 2.73 m. (inner cars).
Gangways: Within unit only.
Wheel Arrangement: 2-Bo – Bo-2 – Bo-2 (– Bo-2) – Bo-2.
Doors: Single-leaf swing plug.

Maximum Speed: 125 m.p.h.
Multiple Working: Classes 220, 221.

DMS. Dia DF201. Bombardier Prorail 2000–01. –/42 1TD 1W.   .   t.
MSRB. Dia. DG201. Bombardier Prorail 2000–01. –/58.   .   t.
MS. Dia. DDG02. Bombardier Prorail 2000–01. –/62 1TD 1W.   .   t.
DMF. Dia DF101. Bombardier Prorail 2000–01. 26/– 1TD 1W.   .   t.

221 001	HX	60331	60851	60951	60751	60451
221 002	HX	60332	60852	60952	60752	60452
221 003	HX	60333	60853	60953	60753	60453
221 004	HX	60334	60854	60954	60754	60454
221 005	HX	60335	60855	60955	60755	60455
221 006	HX	60336	60856	60956	60756	60456
221 007	HX	60337	60857	60957	60757	60457
221 008	HX	60338	60858	60958	60758	60458
221 009	HX	60339	60859	60959	60759	60459
221 010	HX	60340	60860	60960	60760	60460
221 011	HX	60341	60861	60961	60761	60461
221 012	HX	60342	60862	60962	60762	60462
221 013	HX	60343	60863	60963	60763	60463
221 014	HX	60344	60864	60964	60764	60464
221 015	HX	60345	60865	60965	60765	60465
221 016	HX	60346	60866	60966	60766	60466
221 017	HX	60347	60867	60967	60767	60467
221 018	HX	60348	60868	60968	60768	60468
221 019	HX	60349	60869	60969	60769	60469
221 020	HX	60370	60870	60970	60770	60470
221 021	HX	60371	60871	60971	60771	60471
221 022	HX	60372	60872	60972	60772	60472
221 023	HX	60373	60873	60973	60773	60473
221 024	HX	60374	60874	60974	60774	60474
221 025	HX	60375	60875	60975	60775	60475
221 026	HX	60376	60876	60976	60776	60476
221 027	HX	60377	60877	60977	60777	60477
221 028	HX	60378	60878	60978	60778	60478
221 029	HX	60379	60879	60979	60779	60479
221 030	HX	60380	60880	60980	60780	60480
221 031	HX	60381	60881	60981	60781	60481
221 032	HX	60382	60882	60982	60782	60482
221 033	HX	60383	60883	60983	60783	60483
221 034	HX	60384	60884	60984	60784	60484
221 035	HX	60385	60885	60985	60785	60485
221 036	HX	60386	60886	60986	60786	60486
221 037	HX	60387	60887	60987	60787	60487
221 038	HX	60388	60888	60988	60788	60488
221 039	HX	60389	60889	60989	60789	60489
221 040	HX	60390	60890	60990	60790	60490
221 041	HX	60391	60991	60791		60491
221 042	HX	60392	60992	60792		60492
221 043	HX	60393	60993	60793		60493
221 044	HX	60394	60994	60794		60494

3. SERVICE DMUS

This section lists vehicles not used for passenger-carrying purposes. Some vehicles are numbered in the special service stock number series or in the internal user series (An internal user vehicle is a vehicle specifically for use in one location/area which is not otherwise permitted over the Railtrack network without special authority).

CLASS 101 INTERNAL USER OFFICE VEHICLE

Converted 1990 from Class 101 DTC. Gangwayed.
Construction: Steel underframe and aluminium alloy body.
Maximum Speed: 70 m.p.h.
Bogies: DT11.
Brakes: Twin pipe vacuum.
Doors: Manually operated slam.
Couplings: Screw.
Multiple Working: Blue Square.
Dimensions: 18.49 x 2.82 x 3.85 m.
Note: Allocated Internal User number 042222, but this is not carried.

54342. DT. Dia. DZ5??. Lot No. 30468 Metro-Cammell. 1958. 22.5 t.

Spare	**BG**	NS	NL(S)	54342

CLASS 114/1 ROUTE LEARNING UNIT

DMB–DT. Converted 1992 from Class 114/1. Gangwayed within unit.
Construction: Steel.
Engines: Two Leyland TL11/40 of 153 kW (205 h.p.) at 1950 r.p.m. per car.
Transmission: Mechanical. Cardan shaft and freewheel to a four-speed epicyclic gearbox with a further cardan shaft to the final drive, each engine driving the inner axle of one bogie.
Maximum Speed: 70 m.p.h.
Bogies: DD9 + DT9.
Brakes: Twin pipe vacuum.
Doors: Manually operated slam/roller shutter.
Couplings: Screw.
Multiple Working: Blue Square.
Dimensions: 20.45 x 2.82 x 3.87 m.
Non-Standard Livery: Grey, red and yellow.

977775. DMB. Dia. DZ518. Lot No. 30209 Derby 1957. 39.0 t.
977776. DT. Dia. DZ516. Lot No. 30210 Derby 1957. 29.2 t.

-	**0**	E	TE(S)	977775 977776

CLASS 122 ROUTE LEARNING CAR

DM. Converted 1995 from DMBS.
Construction: Steel.
Engines: Two Leyland 1595 of 112 kW (150 h.p.) at 1800 r.p.m.
Gangways: Non gangwayed single car with cab at each end.
Bogies: DD10.
Dimensions: 20.45 x 2.82 x 3.87 m.
Note: Allocated number 977941, but this number is not carried.

55012. DM. Dia. DZ5??. Lot No. 30419 Gloucester 1958. Converted by ABB Doncaster 1995. 36.5 t.

	LH	E	*E*	TE	55012

CLASS 141   WEEDSPRAY UNIT

DM–DM. Built from Leyland National bus components on BREL underframe. 141 105 and 141 112 await conversion.
Construction: Steel.
Engines: One Leyland TL11/65 of 157 kW (210 h.p.) at 1950 r.p.m. per car.
Transmission: Mechanical. SCG R500 4-speed epicyclic gearbox with cardan shafts to SCG RF420i final drive.
Doors: Four-leaf folding.   **Dimensions:** 15.45 x 2.50 x 3.91 m.

55505/12/18. DM. Dia. DZ540. Lot No. 30978 BREL Derby. Converted Serco Railtest Derby 1998. . t.
55525/32/38. DM. Dia. DZ541. Lot No. 30977 BREL Derby. Converted Serco Railtest Derby 1998. . t.

141 105	**WY**	P	*SO*	ZA(S)	55505	55525	
141 112	**WY**	P	*SO*	ZA(S)	55512	55532	
–	**SO**	P	*SO*	ZA	55518	55538	FLOWER

CLASS 930   SANDITE/DE-ICING UNIT

DMB–T–DMB. Converted 1993 from Class 205. Gangwayed within unit. Sandite trailer 977870 is replaced by de-icing trailer 977364 as required.
Construction: Steel.
Engine: One English Electric 4SRKT Mk. 2 of 450 kW (600 h.p.) at 850 r.p.m. per power car.
Transmission: Electric. Two English Electric EE507 traction motors mounted on the inner bogie of each power car.
Maximum Speed: 75 m.p.h.   **Bogies:** SR Mk. 4.
Brakes: Electro-pneumatic and automatic air.
Doors: Manually operated slam.   **Couplings:** Drophead buckeye.
Multiple Working: Classes 201–207.
Dimensions: 20.33 x 2.82 x 3.87 m. (DMB); 20.28 x 2.82 x 3.87 m. (T).

977939–977940. DMB. Dia. DZ537. Lot No. 30671 Eastleigh 1962. 56.0 t.
977870. T. Dia. DZ533. Lot No. 30542 Eastleigh 1960. 30.5 t.

930 301	**RO**	RK	*RK*	SU	977939 977870 977940

CLASS 960   ULTRASONIC TEST TRAIN/TRACTOR UNIT

DM–DM. Converted 1986 from Class 101. Gangwayed within unit. Often operates with either 975091, 999550 or 999602 as a centre car.
Construction: Steel underframe and aluminium alloy body.
Engines: Two Leyland 680/1 of 112 kW (150 h.p.) at 1800 r.p.m. per car.
Transmission: Mechanical. Cardan shaft and freewheel to a four-speed epicyclic gearbox with a further cardan shaft to the final drive, each engine driving the inner axle of one bogie.

Dimensions: 18.49 x 2.82 x 3.85 m.
Maximum Speed: 70 m.p.h.   **Doors:** Manually operated slam.
Bogies: DD15.   **Couplings:** Screw.
Brakes: Twin pipe vacuum.   **Multiple Working:** Blue Square.

977391. DM. Dia. DZ503. Lot No. 30500 Metro-Cammell. 1959. 32.5 t.
977392. DM. Dia. DZ503. Lot No. 30254 Metro-Cammell. 1956. 32.5 t.

-                    **SO**   SO *SO*   RG   977391 977392

CLASS 960   TEST UNIT

DM–DM. Converted 1991 from Class 101. Gangwayed within unit.
Construction: Steel underframe and aluminium alloy body.
Engines: Two Leyland 680/1 of 112 kW (150 h.p.) at 1800 r.p.m. per car.
Transmission: Mechanical. Cardan shaft and freewheel to a four-speed
epicyclic gearbox with a further cardan shaft to the final drive, each engine
driving the inner axle of one bogie.
Maximum Speed: 70 m.p.h.
Bogies: DD15.   **Couplings:** Screw.
Brakes: Twin pipe vacuum.   **Multiple Working:** Blue Square.
Doors: Manually operated slam.   **Dimensions:** 18.49 x 2.82 x 3.85 m.

977693. DM. Dia. DZ503. Lot No. 30261 Metro-Cammell. 1957. 32.5 t.
977694. DM. Dia. DZ503. Lot No. 30276 Metro-Cammell. 1958. 32.5 t.

-                    **SO**   SO *SO*   BY   977693 977694 Iris 2

CLASS 960   SANDITE UNIT

DMB. Converted 1991/93 from Class 121. Non gangwayed.
Construction: Steel.
Engines: Two Leyland 1595 of 112 kW (150 h.p.) at 1800 r.p.m.
Transmission: Mechanical. Cardan shaft and freewheel to a four-speed
epicyclic gearbox with a further cardan shaft to the final drive, each engine
driving the inner axle of one bogie.
Maximum Speed: 70 m.p.h.
Bogies: DD10.   **Couplings:** Screw.
Brakes: Twin pipe vacuum.   **Multiple Working:** Blue Square.
Doors: Manually operated slam.   **Dimensions:** 20.45 x 2.82 x 3.87 m.

977722-977723. DMB. Dia. DZ515. Lot No. 30518 Pressed Steel 1960. 38.0 t.
977858–60/66/73. DMB. Dia. DZ526. Lot No. 30518 Pressed Steel 1960. 38.0 t.

960 002	**N**	RK		AL(S)	977722
121 021	**RO**	RK	*RK*	AL	977723
55024	**M**	RK	*RK*	AL	977858
960 011	**RK**	RK		AF	977859
960 012	**N**	RK	*RK*	AF	977860
960 013	**RO**	RK		RG(S)	977866
960 014	**N**	RK		RG(S)	977873

CLASS 960　　　　　　　　　　　SANDITE UNIT

DMB. Converted 1991 from Class 122 vehicle. Non gangwayed.
Construction: Steel.
Engines: Two Leyland 1595 of 112 kW (150 h.p.) at 1800 r.p.m.
Transmission: Mechanical. Cardan shaft and freewheel to a four-speed epicyclic gearbox with a further cardan shaft to the final drive, each engine driving the inner axle of one bogie.　　**Maximum Speed:** 70 m.p.h.
Bogies: DD10.　　　　　　　　　　　**Couplings:** Screw.
Brakes: Twin pipe vacuum.　　　　**Multiple Working:** Blue Square.
Doors: Manually operated slam.　　**Dimensions:** 20.45 x 2.82 x 3.87 m.

975042. DM. Dia. DX516. Lot No. 30419 Gloucester 1958. 36.5 t.

122 019		**RO**	RK *RK*	AL	975042

CLASS 960/9　　SANDITE & ROUTE LEARNING UNIT

DM–DM. Converted 1993 from Class 101 vehicles. Gangwayed within unit.
Construction: Steel underframe and aluminium alloy body.
Engines: Two Leyland 680/1 of 112 kW (150 h.p.) at 1800 r.p.m. per car.
Transmission: Mechanical. Cardan shaft and freewheel to a four-speed epicyclic gearbox with a further cardan shaft to the final drive, each engine driving the inner axle of one bogie.　　**Maximum Speed:** 70 m.p.h.
Bogies: DD15.　　　　　　　　　　　**Couplings:** Screw.
Brakes: Twin pipe vacuum.　　　　**Multiple Working:** Blue Square.
Doors: Manually operated slam.　　**Dimensions:** 18.49 x 2.82 x 3.85 m.

977895. DM. Dia. DZ503. Lot No. 30275 Metro-Cammell. 1958. 32.5 t.
977896/900. DM. Dia. DZ504. Lot No. 30276 Metro-Cammell. 1958. 32.5 t.
977897/901/903. DM. Dia. DZ503. Lot No. 30259 Metro-Cammell. 1957. 32.5 t.
977898. DM. Dia. DZ515. Lot No. 30256 Metro-Cammell. 1957. 32.5 t.
977899. DM. Dia. DZ503. Lot No. 30500 Metro-Cammell. 1959. 32.5 t.
977902. DM. Dia. DZ503. Lot No. 30261 Metro-Cammell. 1957. 32.5 t.
977904. DM. Dia. DZ503. Lot No. 30270 Metro-Cammell. 1957. 32.5 t.

960 991	**N**	RK	LO(S)	977895 977896
960 992	**BG**	RK	LO(S)	977897 977898
960 993	**BG**	RK	LO(S)	977899 977900
960 994	**BG**	RK	LO(S)	977901 977902
960 995	**BG**	RK	LO(S)	977903 977904

UNCLASSIFIED　　　　　　　　　DE-ICING CAR

T. Converted 1960 from 4-Sub EMU vehicle. Non gangwayed. Operates with 977939/40.
Construction: Steel.
Maximum Speed: 70 m.p.h.　　　　**Couplings:** Drophead buckeye.
Bogies: Central 43 inch.　　　　　**Multiple Working:** SR system.
Brakes: Electro-pneumatic and automatic air.

Doors: Manually operated slam. **Dimensions:**

977364. T. Dia. EZ520. Eastleigh 1946. 29.0 t.

| - | **RO** | RK | *RK* | SU | 977364 |

UNCLASSIFIED   TRACK RECORDING CAR

T. Purpose built service vehicle. Wired for DMU working 1998. Gangwayed.
Operates with 977391/2 or loco-hauled.

Construction: Steel.   **Maximum Speed:** 70 m.p.h.
Bogies: B4.   **Couplings:** Drophead buckeye.
Brakes: Air and vacuum.   **Multiple Working:** Blue square.
Doors: Manually operated slam.   **Dimensions:**

999550. T. Dia. DZ539. Lot No. 3830 BREL Derby 1976. 45.0 t.

| - | **SO** | SO | *SO* | ZA | 999550 |

UNCLASSIFIED   TRACK ASSESSMENT UNIT

DM–DM. Purpose built service unit. Gangwayed within unit.
Construction: Steel.
Engine: One Cummins NT-855-RT5 of 213 kW (285 h.p.) at 2100 r.p.m. per power car.
Transmission: Hydraulic. Voith T211r with cardan shafts to Gmeinder GM190 final drive.
Maximum Speed: 75 m.p.h.   **Couplers:** BSI automatic.
Bogies: BP38 (powered), BT38 (non-powered).
Brakes: Electro-pneumatic.   **Dimensions:** 20.06 x 2.82 x 3.77 m.
Doors: Manually operated slam & power operated sliding.
Multiple Working: With classes 141–158 and 170 only.
Non-Standard Livery: Grey, red and blue.

999600. DM. Dia. DZ536. Lot No. 4060 BREL York 1987. 36.5 t.
999601. DM. Dia. DZ536. Lot No. 4061 BREL York 1987. 36.5 t.

| - | **0** | SO | *SO* | NC | 999600 999601 |

UNCLASSIFIED   ULTRASONIC TEST CAR

T. Converted 1986 from Class 432 EMU. Gangwayed. Operates with 977391/2.
Construction: Steel.   **Maximum Speed:** 70 m.p.h.
Bogies: SR Mk. 6.   **Couplings:** Screw.
Brakes: Twin pipe vacuum.   **Multiple Working:** Blue Square.
Doors: Manually operated slam.   **Dimensions:** 19.66 x 2.82 x 3.90 m.

999602. T. Dia. DZ531. Lot No. 30862 York 1974. 55.5 t.

| - | **SO** | SO | *SO* | ZA | 999602 |

▲ Class 165 'Turbo' No. 165 127 approaches Bath Spa with the 13.40 Oxford–Bristol Temple Meads service on 5th May 1999. The Network SouthEast livery, still carried by this unit, is now obsolescent. **John Chalcraft**

▼ The recently introduced Thames Trains livery is seen here on Class 166 No. 166 201 as it pauses at Reading whilst forming the 15.15 Oxford–London Paddington service on 29th September 2000. **D. Ford**

▲ Chiltern Railways liveried Class 168/1 No. 168 109 passes West Ruislip with the 11.30 Birmingham Snow Hill–London Marylebone 'Clubman' service. The outer cars of these units are, in reality, Class 170s with Class 168 interiors.

David Brown

▼ Class 170 No. 170 102 passes through Cricklewood with the 11.56 Nottingham–London St Pancras on 7th April 2000. This unit carries Midland Mainline livery. **K. Conkey**

▲ Anglia Railways liveried Class 170 No. 170 208 forms the 12.32 Basingstoke–Chelmsford at Winchfield on 19th July 2000. This is one of a number Basingstoke to East Anglia services recently introduced by Anglia Railways.   **David Brown**

▼ Central Trains liveried Class 170 No. 170 636 nears the end of its journey as it passes Alexandra Dock Junction, west of Newport, with the 10.28 Nottingham–Cardiff Central on 30th June 2000.   **Rodney Lissenden**

▲ After much delay, units of Class 175 are now entering service with First North Western. Carrying First Group livery, No. 175 006 is seen here near Stableford, Staffordshire whilst working a Birmingham to Holyhead service. The date is 19th July 2000. **Hugh Ballantyne**

▼ Connex South Central liveried Class 205 No. 205 009 stables between duties at Selhurst T&RSMD on 16th November 1999. **Brian Denton**

▲ Testing of the first Class 220 Virgin 'Voyager' DEMU has recently commenced in Belgium, where the class are under construction. The first unit is pictured here between trials on the Belgian network. **Daniel Moens**

▼ Loadhaul liveried Class 122 'bubble' car, No. 55012, is used on route learning duties. In this capacity, it was pictured at Brocklesby on 7th July 1999.
Ian A. Lyall

Croydon Tramlink. Carrying 'Necafé' advertising livery, car No. 2533 passes through East Croydon whilst working a Route 2 service from Croydon–Beckenham Junction.

Peter Fox

Manchester Metrolink. One of the recently built Ansaldo cars, No. 2002 is pictured on Eccles New Road on 22nd July 2000, the first day of normal service on the Eccles extension.

Peter Fox

▲ **Strathclyde PTE Underground.** Cars Nos. 105, 206 and 128 are pictured in a 3-car formation outside the depot at Broomloan on 15th May 1999. **Ross Aitken**

▼ **Tyne and Wear Metro.** Two units pass at West Jesmond on 8th April 2000. On the left, unit No. 4012 carries red and yellow livery and is working a Newcastle Airport bound service, whilst the unit on the right, No. 4002, is in advertising livery and working a service destined for South Shields. **Rodney Lissenden**

4. VEHICLES AWAITING DISPOSAL

Class 100

977191	**B**	ZC	

Class 101

101 657	**RR**	A	LO	53211 54085
101 660	**RR**	A	PY	51189 54343 51213
101 665	**RR**	A	PY	51429 54393
101 686	**S**	A	PY	51231 51500
101 690	**S**	A	PY	51435 53177
L835	**RR**	A	PY	51432 51498
L840	**N**	A	LO	53311 53322
L842	**N**	A	ZA	53314 53327
Spare	**RR**	A	LO	51463
Spare:	**RR**	A	PY	54055 54061 54352 54365
Spare	**RR**	A	LO	54062 54091
Spare:	**BG**	A	PY	54350
Spare	**RR**	A	BP	59303
Spare	**G**	A	BP	59539

Class 117

117 301	**RR**	A	PY	51353	51395	L704	**N**	A	PY	51341	51383
117 306	**RR**	A	AL	51369	51411	L706	**N**	A	PY	51366	51408
117 308	**RR**	A	PY	51371	51413	L707	**N**	A	PY	51335	51377
117 310	**RR**	A	PY	51373 59486 51381		L720	**N**	A	PY	51354	51396
117 311	**RR**	A	PY	51352	51376	L721	**N**	A	PY	51363	51405
117 313	**RR**	A	PY	51339	51382	Spare	**RR**	A	PY	59492 59500 59505
117 701	**N**	A	PY	51350	51392	Spare	**RR**	A	PY	59509 59521
L702	**N**	A	PY	51356	51398	Spare	**N**	A	PY	51358	51400

Class 117

| 121 027 | **SL** | A | PY | 55027 | | 121 031 | **N** | A | PY | 55031 |
| 121 029 | **SL** | A | PY | 55029 | | | | | | |

Class 141

141 101	**WY**	P	ZF	55501 55521		141 114	**WY**	CW	ZF	55514 55534
141 102	**WY**	CW	ZF	55502 55522		141 115	**WY**	CW	ZF	55515 55535
141 106	**WY**	CW	ZF	55506 55526		141 116	**WY**	CW	MM	55516 55536
141 107	**WY**	CW	ZF	55507 55527		141 117	**WY**	CW	HT	55517 55537
141 109	**WY**	CW	ZF	55509 55529		141 119	**WY**	CW	ZF	55519 55539
141 111	**WY**	CW	ZF	55511 55531		141 120	**WY**	CW	ZF	55520 55540
141 113	**WY**	P	ZF	55513 55533						

Class 210

| 210 001 | **N** | CW | ZG | 60200 60201 |

Class 951

| 977696 | **N** | | ZG | |

5. UK LIGHT RAIL SYSTEMS & METROS

5.1. BLACKPOOL & FLEETWOOD TRAMWAY

Until the opening of Manchester Metrolink, the Blackpool & Fleetwood Tramway was the only urban/inter-urban tramway system left in Britain. The infrasctructure is owned by the local authorities and the tramway operated by Blackpool Transport Services Ltd.

System: 660 V d.c. overhead.
Depot & Workshops: Rigby Road, Blackpool.
Livery: Cream and green except where stated otherwise.

Notes: Numbers in brackets are pre-1968 numbers. All cars are single-deck unless stated otherwise.

OPEN BOAT CARS   A1-1A

Built: 1934–5 by English Electric. 12 built (225–236).
Traction Motors: Two EE327 of 30 kW.
Seats: 56.

600 (225)	604 (230)	607 (236)
602 (227)	605 (233)	

REPLICA VANGUARD   A1-1A

Built: 1987 by Blackpool & Fleetwood Tramway, Blackpool on underframe of one man car No. 7 which was itself converted from English Electric Railcoach 619 (282).
Traction Motors: Two EE327 of 30 kW.
Seats:

619

BRUSH RAILCOACHES   A1-1A

Built: 1937 by Brush. 20 built (284–303).
Traction Motors: Two EE305 of 40 kW (*EE327 of 30 kW).
Seats: 48.

Note: 635 carries its original number 298.

621	(284)		**AL**	627	(290)	**Purple**	
622	(285)	*	**White**	630	(293)	**AL**	
623	(286)		**AL**	631	(294)		
625	(288)			632	(295)	**AL**	
626	(289)		**AL**				

633	(296)		
634	(297)	**AL**	
636	(299)		
637	(300)	**AL**	

CENTENARY CLASS A1-1A

Built: 1984–7. Body by East Lancs. Coachbuilders, Blackburn. One man operated.
Traction Motors: Two EE305 of 40 kW.
Seats: 52.

* Rebuilt from GEC car 651.

641	**AL**	643	**AL**	645	**AL**	647	**AL**
642	**AL**	644	**AL**	646	**AL**	648 *	

CORONATION CLASS Bo-Bo

Built: 1953 by Charles Roberts & Co., Wakefield. Resilient wheels. 25 built (304–328).
Traction motors: Four Crompton-Parkinson 92 of 34 kW.
Seats: 56.

660 (324)

PROGRESS TWIN CARS A1-1A + 2-2

Built: Motor cars (671–677) rebuilt 1958–60 from English Electric railcoaches by Blackpool Corporation Transport. Driving trailers (681–687) built 1960 by Metro-Cammell.
Traction Motors: Two EE305 of 40 kW.
Seats: 53 + 53.

671+681	(281+T1)	674+684	(284+T4)	676+686	(286+T6)
672+682	(282+T2)	675+685	(285+T5)	677+687	(287+T7)
673+683	(283+T1)				

ENGLISH ELECTRIC RAILCOACHES A1-1A

Built: Rebuilt 1958–60 from English Electric railcoaches. Originally ran with trailers.
Traction Motors: Two EE305 of 40 kW.
Seats: 48.

678 (278) **AL** | 679 (279) | 680 (280) **AL**

"BALLOON" DOUBLE DECKERS A1-1A

Built: 1934–5 by English Electric. 700–712 were originally built with open tops, and 706 has now reverted to that condition and is named 'PRINCESS ALICE'.
Traction Motors: Two EE305 of 40 kW.
Seats: 94 (*† 92, ‡ 90, ¶ 88).

* Rebuilt with new front end design and air-conditioned cabs. Known as "Super Balloons" or "Millennium Class"

§ Converted to ice cream tram seating 78 with an ice cream sales area in one of the lower saloons.

o Rebuilt as open-topped double-decker setaing 82. Named "PRINCESS ALICE'

700	(237)		709	(246)	*	718	(255)		
701	(238)	‡	710	(247)		719	(256)	§	**AL**
702	(239)	¶	711	(248)	†	720	(257)		**AL**
703	(240)		712	(249)		721	(258)		**AL**
704	(241)	**AL**	713	(250)		722	(259)	‡	**AL**
706	(243)	o	715	(252)		723	(260)	†	**AL**
707	(244)	*	716	(253)		724	(261)		
708	(245)	‡	717	(254)		726	(263)		**AL**

ILLUMINATED CARS

732	(168)	Rocket	Built: 1961	Seats: 47
733	(209)	Western Train loco. & tender	Built: 1962	Seats: 35
734	(174)	Western Train coach	Built: 1962	Seats: 60
735	(222)	Hovertram	Built: 1963	Seats: 99
736	(170)	HMS Blackpool	Built: 1965	Seats: 71

JUBILEE CLASS DOUBLE DECKERS

Built: Rebuilt 1979/82 from Balloon cars. Standard bus ends, thyristor control and stairs at each end. 761 has one door per side whereas 762 has two.
Traction Motors: Two EE305 of 40 kW.
Seats: 100.

761   (725, 262) **AL** | 762   (714, 251)

WORKS CARS

259	(748, 624)	Permanent way car	Converted: 1971
260	(751, 628, 291)	Crane car and rail carrier	Converted: 1973
752	(2, 1)	Rail grinder and snowplough	Converted: 19
754		New works car (unnumbered)	Built: 1993

VINTAGE CARS

Stockport 5	Double-decker	Built: 1901
Bolton 66	Bogie double-decker	Built: 1901
Blackpool & Fleetwood 40	Box car	Built: 1914
Blackpool & Fleetwood 147	Standard double-decker	Built: 1924
Glasgow 1245	Double-decker	Built: 1939

5.2 CROYDON TRAMLINK

This new system runs through central Croydon with lines to Wimbledon, Addington and Beckenham Junction/Elmers End. Operated by Tramlink Croydon Ltd.

System: 750 V d.c. overhead.
Depot & Workshops: Therapia Lane, Croydon.
Livery: Red and white.

SIX AXLE ARTICULATED CARS   Bo–2–Bo

Built: 1998-99 by Bombardier-Wien Schienenfahrzeuge, Austria.
Traction Motors: Four of 120 kW each.
Seats: 70.
Dimensions: 30.10 x 2.65 m.
Couplers: Scharfenberg.
Doors: Sliding plug.
Weight: 36.3 t.
Braking: Disc, regenerative and magnetic track.
Max. Speed: 80 km/h.

The following trams carry advertising livery:
* 2531 Addington Palace Country Club (purple)
* 2533 Nescafé (dark brown)
* 2542 Amey Construction (yellow)
* 2546 Whitgift Shopping Centre (blue)
* 2550 First Group (similar to **FN** – indigo blue with pink and white stripes).

2530	2534	2538	2542 **AL**	2546 **AL**	2550 **AL**
2531 **AL**	2535	2539	2543	2547	2551
2532	2536	2540	2544	2548	2552
2533 **AL**	2537	2541	2545	2549	2553

5.3. DOCKLANDS LIGHT RAILWAY

This is a light rail line running in London's East End between Bank, Tower Gateway and Stratford to Lewisham and Beckton. Originally owned by London Transport, it is now owned by DLR Ltd. and operated by DLR Management Ltd. Cars normally operate automatically using the Alcatel "Seltrack" moving block signalling system.

System: 750 V d.c. third rail (bottom contact).
Depots: Poplar, Beckton.
Workshops: Poplar.
Livery: Red and blue.

CLASS B90 B–2–B

Built: 1991–2 by BN Construction, Brugge, Belgium. Chopper control.
Traction Motors: Two Brush of 140 kW.
Seats: 66.   **Weight:** 36 t.
Dimensions: 28.80 x 2.65 m.   **Braking:** Rheostatic.
Couplers: Scharfenberg.   **Max. Speed:** 80 km/h.
Doors: Sliding. End doors for staff use.

22	26	30	34	38	42
23	27	31	35	39	43
24	28	32	36	40	44
25	29	33	37	41	

CLASS B92 B–2–B

Built: 1992–5 by BN Construction, Brugge, Belgium. Chopper control.
Traction Motors: Two Brush of 140 kW.
Seats: 66.   **Weight:** 36 t.
Dimensions: 28.80 x 2.65 m.   **Braking:** Rheostatic.
Couplers: Scharfenberg.   **Max. Speed:** 80 km/h.
Doors: Sliding. End doors for staff use.

45 **AL**	53	61	69	77	85
46 **AL**	54	62	70	78	86
47 **AL**	55	63	71	79	87
48 **AL**	56	64	72 **AL**	80	88
49 **AL**	57	65	73	81	89
50 **AL**	58	66	74	82	90
51 **AL**	59	67 **AL**	75	83	91
52	60	68 **AL**	76	84	

CLASS ? B–2–B

To be built: by Bombardier Transportation, Brugge, Belgium.
Traction Motors: Two Brush of 140 kW.
Seats: 66.   **Weight:** 36 t.
Dimensions: 28.80 x 2.65 m.   **Braking:** Rheostatic.

Couplers: Scharfenberg. **Max. Speed:** 80 km/h.
Doors: Sliding. End doors for staff use.

92	96	100	104	108	112
93	97	101	105	109	113
94	98	102	106	110	114
95	99	103	107	111	115

BATTERY/ELECTRIC WORKS LOCO B

Built: 1991 by RFS Engineering, Kilnhurst.
Electrical Equipment:

Unnumbered

DIESEL SHUNTER B

Built: 1962 by Ruston & Hornsby, Lincoln.
Engine:
Transmission: Mechanical.

Unnumbered

5.4. GREATER MANCHESTER METROLINK

This light rail system runs from Bury to Altrincham through the streets of Manchester, with a spur to Piccadilly. A new line has now opened to Salford Quays and Eccles. The system is franchised to the Altram consortium and the operation is subcontracted to Serco Metrolink.

System: 750 V d.c. overhead.
Depot & Workshops: Queens Road, Manchester.
Livery: White, dark grey and blue.

SIX-AXLE ARTICULATED CARS Bo–2–Bo

Built: 1991–2 by Firema, Italy. Chopper control.
Traction Motors: Four GEC of 130 kW.
Seats: 84.
Dimensions: 29.00 x 2.65 m.
Couplers: Scharfenberg.
Doors: Sliding.
Weight: 45 t.
Braking: Rheostatic, regenerative, disc and emergency track.
Max. Speed: 80 km/h.

* Fitted with front valances, retractible couplers and controllable magnetic track brakes for running to Eccles.

1001	
1002	
1003	
1004	THE ROBERT OWEN
1005	*

1006		
1007		
1008		MANCHESTER AIRPORT
1009		
1010	*	MANCHESTER CHAMPION
1011		
1012		
1013		THE FUSILIER
1014		THE CITY OF DRAMA
1015	*	SPARKY
1016		
1017		
1018		SIR MATT BUSBY
1019		
1020		THE DAVID GRAHAM CBE
1021		THE GREATER MANCHESTER RADIO
1022		THE GRAHAM ASHWORTH
1023		
1024		THE JOHN GREENWOOD
1025		
1026		THE POWER

SIX-AXLE ARTICULATED CARS   Bo–2–Bo

Built: 1999 by Ansaldo, Italy. Chopper control. Fitted with front valances, retractible couplers and controllable magnetic track brakes for running to Eccles.
Traction Motors: Four GEC of 130 kW.
Seats: 82.
Dimensions: 29.00 x 2.65 m.
Couplers: Scharfenberg.
Doors: Power operated sliding.
Weight: 45 t.
Braking: Rheostatic, regenerative, disc and magnetic track.
Max. Speed: 80 km/h.

2001	
2002	
2003	TRAVELLER 2000
2004	
2005	
2006	CITY OF SALFORD 2000

SPECIAL PURPOSE VEHICLE

Built: 1991 by RFS Industries, Kilnhurst.
Engine: Caterpillar 3306 PCT of 170 kW.
Transmission: Mechanical. Rockwell T280.
Couplers: Scharfenberg.
Max. Speed: 40 km/h.

Unnumbered

5.5 MIDLAND METRO

This new operation runs from Birmingham Snow Hill to Wolverhampton along the former GWR line to Wolverhampton Low Level. On the approach to Wolverhampton it deviates from the former railway alignment to run on-street to the St. George's terminus. Operated by Travel West Midlands Ltd.

System: 750 V d.c. overhead.
Depot & Workshops: Wednesbury.
Livery: Dark blue and light grey with green stripe, yellow doors and red front end and roof.

SIX AXLE ARTICULATED CARS   Bo–2–Bo

Built: 1998–99 by Ansaldo Transporti, Italy.
Traction Motors: Four.
Seats: 58.
Dimensions: 24.00 x 2.65 m.
Couplers: Not equipped.
Doors: Power operated sliding plug.
Weight: 35.6 t.
Braking: Rheostatic, regenerative, disc and magnetic track.
Max. Speed: 75 km/h.

01	04	07	10	13	15
02	05	08	11	14	16
03	06	09	12		

5.6. STAGECOACH SUPERTRAM

This light rail system has three lines, to Halfway in the south east of Sheffield with a spur from Gleadless Townend to Herdings, to Middlewood in the north west with a spur from Hillsborough to Malin Bridge and to Meadowhall Interchange in the north east adjacent to the large shopping complex. Because of the severe gradients in Sheffield (up to 1 in 10), all axles are powered on the vehicles which are owned by South Yorkshire Light Rail Ltd., a subsidiary of South Yorkshire Passenger Transport Executive, but are mortgaged to Lloyd's Bank, whilst the operating company, South Yorkshire Supertram Ltd. has been leased to Stagecoach Holdings Ltd. for 27 years and is now known as Stagecoach Supertram.

System: 750 V d.c. overhead.
Depot & Workshops: Nunnery.
Livery: White with orange, red and blue stripes.

AL Advertising livery (JON)

EIGHT-AXLE ARTICULATED UNITS   B–B–B–B

Built: 1993–4 by Duewag, Düsseldorf, Germany.
Traction Motors: Four monomotors.
Seats: 88.
Dimensions: 34.75 x 2.65 m.
Couplers: Not equipped.
Doors: Sliding plug.
Weight: 52 t.
Braking: Rheostatic, regenerative, disc and emergency track.
Max. Speed: 50 m.p.h.

101	106	110	114	118	122
102	107	111	115	119	123
103	108	112	116	120 **AL**	124
104	109	113	117	121	125
105					

5.7. STRATHCLYDE PTE UNDERGROUND

This circular 4 foot gauge underground line in Glasgow is generally referred to as the "Subway" or the "Clockwork Orange". It is operated by Strathclyde PTE.

System: 600 V d.c. third rail.
Depot & Workshops: Broomloan.
Livery: Orange and black.

SINGLE POWER CARS   Bo–Bo

Built: 1977–79 by Metropolitan Cammell, Birmingham. Refurbished 1993–95 by ABB Derby.
Traction Motors: Four GEC G312AZ of 35.6 kW each.
Seats: 36.
Dimensions: 12.81 x 2.34 m.
Couplers: Wedglock.
Doors: Sliding.
Weight: 19.62 t.
Maximum Speed: 54 km/h.

101	107	113	119	124	129
102	108	114	120	125	130
103	109	115	121	126	131
104	110	116	122	127	132
105	111	117	123	128	133
106	112	118			

INTERMEDIATE TRAILERS   2–2

Built: 1992 by Hunslet-Barclay, Kilmarnock.
Seats: 40.
Dimensions: 12.70 x 2.34 m.
Couplers: Wedglock.
Doors: Sliding.
Weight: 17.25 t.
Maximum Speed: 54 km/h.

201	203	205	206	207	208
202	204				

BATTERY ELECTRIC WORKS LOCOS   Bo

Built: 1977 by Clayton Equipment, Hatton.
Battery: 96 cell Chloride lead acid.
Traction Motors: Two GEC G312AZ of 35.6 kW each.
Dimensions: 4.65 x ?.?? m.
Couplers: Wedglock.

Weight: 14.8 tonnes.
Braking:
Max. Speed:

L2   LOBEY DOSSER
L3   RANK BAJIN

BATTERY ELECTRIC WORKS LOCO   Bo

Built: 1974 by Clayton Equipment, Hatton. One of a pair of 3 ft. gauge locomotives for use on the Channel Tunnel construction project. Both were converted to 4 ft. gauge in 1976 or 1977 and used by contractors Taylor Woodrow on the Strathclyde Underground modernisation between 1977 and 1980 before being stored at Taylor Woodrow, Southall. Both were purchased in a derelict condition by Strathclyde PTE in 1987, and one loco, L4, was rebuilt by Clayton in 1988 from the parts recovered from these two locomotives. L4 was further rebuilt by Hunslet-Barclay, Kilmarnock, in 1990 to make it compatible with L2 and L3.
Battery: 96 cell Chloride lead acid.
Traction Motors: Two GEC G312AZ of 35.6 kW each.
Dimensions:
Couplers: Wedglock.
Weight:
Max. Speed:

L4   EL FIDELDO

5.8. TYNE AND WEAR METRO

System: 1500 V d.c. overhead.
Depot & Workshops: South Gosforth.

SIX-AXLE ARTICULATED UNITS   B–2–B

Built: 1978–81 by Metropolitan Cammell, Birmingham (4001/2 were built by Metropolitan Cammell in 1976 and rebuilt 1984–87 by Hunslet TPL, Leeds).
Traction Motors: Two Siemens of 187 kW each.
Seats: 68.   **Dimensions:** 27.80 x 2.65 m.
Couplers: BSI.   **Doors:** Sliding plug.
Weight: 39.0 t.   **Maximum Speed:** 80 km/h.
Livery: Red &Yellow unless otherwise indicated.

AL Advertising livery; **B** Blue and yellow; **G** Green and yellow.
0 Original 1975 livery with red stripe as used on test track.

4001	**0**	4019		4037		4055	**AL**	4073	
4002	**AL**	4020		4038		4056	**G**	4074	
4003		4021		4039	**AL**	4057		4075	**B**
4004	**G**	4022		4040		4058	**B**	4076	**B**
4005		4023	**G**	4041		4059		4077	
4006		4024		4042	**AL**	4060		4078	
4007		4025	**G**	4043		4061	**G**	4079	
4008	**B**	4026		4044		4062	**G**	4080	
4009		4027		4045	**AL**	4063	**AL**	4081	**B**
4010		4028		4046		4064		4082	**G**
4011		4029		4047	**B**	4065		4083	**B**
4012		4030		4048	**B**	4066	**B**	4084	
4013		4031	**B**	4049	**AL**	4067		4085	**B**
4014		4032		4050		4068		4086	**AL**
4015		4033		4051		4069		4087	**AL**
4016	**B**	4034		4052		4070		4088	
4017		4035	**B**	4053	**B**	4071		4089	
4018	**G**	4036	**G**	4054	**B**	4072	**G**	4090	

Names:

4026	George Stephenson	4065	Dame Catherine Cookson
4041	HARRY COWANS	4077	Robert Stephenson
4060	Thomas Bewick	4078	ELLEN WILKINSON

BATTERY/ELECTRIC WORKS LOCOS

Built: 1989–90 by Hunslet TPL, Leeds.
Traction Motors: Two Hunslet-Greenbat T9-4P of 67 kW each.
Dimensions: 9.00 x ?.?? m.   **Couplers:** BSI.
Weight: 26.25 t.   **Maximum Speed:** 50 km/h.
Livery: Green, with red solebar and black & yellow striped ends.

BL1 | BL2 | BL3

6. CODES

LIVERY CODES

Code	Description
AL	Advertising livery (See class heading for details).
AN	Anglia Railways (White/turquoise with blue vignette).
AR	Anglia Railways (Turquoise blue with white stripe).
BG	BR (Blue and grey lined out in white).
CO	Centro (Grey/green with light blue, white & yellow stripes).
CR	Chiltern Railways (Blue and white with a thin red stripe).
CT	Central Trains (Two-tone green with yellow doors. Blue flash at vehicle ends. Thin red stripe at vehicle ends and at cantrail level).
CX	Connex (Light grey with yellow lower body vignette and blue solebar).
FN	First North Western (Indigo blue with pink and white stripes).
FW	First Great Western (Indigo blue with white roof and gold, pink and white stripes).
G*	BR (Green with straw stripe).
GM	Greater Manchester PTE (Light grey/dark grey with red and white stripes).
LH*	BR Loadhaul (Black with orange cab sides).
M	BR Maroon (Maroon lined out in straw and black).
MM	Midland Mainline (Teal green with cream lower body sides and three orange stripes).
MT	Merseytravel (Yellow/white with grey and black stripes).
MY	Merseytravel (Yellow/white with grey stripe).
N	Network South East (Grey/white/red/white/blue/white).
NS	Northern Spirit (Turquoise blue with lime green 'N').
NT	Network South East (Grey/red/white/blue/white).
NW	North West Trains (Blue with gold cant rail stripe and star).
O	Non-standard livery (see class heading for details)
PS	Provincial (Dark blue/grey with light blue & white stripes).
RE*	Provincial Express (Light grey/buff/dark grey with white, dark blue & light blue stripes).
RK*	Railtrack (Orange with white and grey stripes).
RN*	North West Regional Railways (Dark blue/grey with green & white stripes).
RR*	Regional Railways (Dark Blue/Grey with light blue & white stripes).
S	Strathclyde PTE (Orange/black lined out in white).
SP	Strathclyde PTE (Carmine & cream lined out in black and gold).
SL	Silverlink (Indigo blue with white stripe, green lower body and yellow doors).
SO	Serco Railtest (Red/grey).
SR	ScotRail (White, terracotta, purple and aquamarine).
SW	South West Trains (White/black/orange/red/blue with red doors).
TW	Tyne & Wear PTE (White/yellow with blue stripe).
TX	Northern Spirit Trans-Pennine Express (Plum with yellow 'N').

VL Valley Lines. (Dark green & red with white & light green stripes. Light green doors).

VT Virgin Trains (Silver, with black window surrounds, white cant-rail stripe and red roof. Red is swept down at unit ends. Black and white striped doors).

WW Wales & West Passenger Trains (Grey, orange, light blue and dark blue with orange doors)

WY* West Yorkshire PTE (Red/cream with thin yellow stripe).

YN West Yorkshire PTE (Red with light grey 'N').

* denotes an obsolescent livery style no longer used for repaints.

OWNER CODES

A   Angel Train Contracts
BB   Bridgend County Borough Council
CW   Cotswold Rail Ltd.
E   English Welsh & Scottish Railway
H   HSBC Rail (UK) Ltd.
HD   Hastings Diesels Ltd.
HX   Halifax Asset Finance Ltd.
NS   Northern Spirit
P   Porterbrook Leasing Company
RD   Rhondda Cynon Taff District Council
RI   Rail Assets Investments Ltd.
RK   Railtrack
SO   Serco Railtest
W   Wiltshire Leasing Co. (First Group)

OPERATION CODES

AR   Anglia Railways.
CA   Cardiff Railway Company (NB: All units used must be equipped global positioning system).
CR   Chiltern Railways.
CT   Central Trains.
E   English Welsh & Scottish Railway.
GW   First Great Western.
MM   Midland Mainline.
NS   Northern Spirit.
NW   First North Western.
ON   Normally used on special or charter services.
RK   Railtrack.
SC   Connex South Central.
SL   Silverlink.
SO   Serco Railtest.
SR   ScotRail.
SW   South West Trains.
TT   Thames Trains.

VX	Virgin Cross Country
VX	Virgin West Coast
WW	Wales and West Passenger Trains

DEPOT, WORKS & LOCATION CODES

* denotes unofficial code.

Code	Location	Operator
AF	Chart Leacon (Ashford, Kent) T&RSMD	Adtranz
AL	Aylesbury TMD	Chiltern Railways
BP	Blackpool North Carrige Sidings	*Storage location only*
BY	Bletchley T&RSMD	Silverlink
CF	Cardiff Canton TMD	Wales & West
CH	Chester TMD	First North Western
CK	Corkerhill (Glasgow) TMD	Scotrail
DY	Derby Etches Park T&RSMD	Maintrain
HA	Haymarket (Edinburgh) TMD	Scotrail
HT	Heaton T&RSMD (Newcastle)	Northern Spirit
KR	Kidderminster	Severn Valley Railway
LO	Longsight TMD (D) (Manchester)	First North Western
MM*	Fire Service College, Moreton-in-Marsh	*Storage location only*
NC	Norwich Crown Point T&RSMD	Anglia Railways
NH	Newton Heath T&RSMD (Manchester)	First North Western
NL	Neville Hill DMU/EMU (Leeds) T&RSMD	Northern Spirit
OM	Old Oak Common (London) CARMD	First Great Western
PY*	MoD Pig's Bay (Shoeburyness)	*Storage location only*
RG	Reading TMD	Thames Trains
SA	Salisbury TMD	South West Trains
SE	St. Leonards T&RSMD	St. Leonards Railway Engineering.
SU	Selhurst (Croydon) T&RSMD	Connex South Central
TE	Thornaby T&RSMD	EWS
TS	Tyseley (Birmingham) T&RSMD	Maintrain
ZA	Railway Technical Centre, Derby	Serco Railtest/ Fragonset Railways
ZC	Crewe	Adtranz
ZD	Litchurch Lane, Derby	Adtranz
ZF	Doncaster	Adtranz
ZG	Eastleigh	Alstom
ZH	Springburn	Railcare
ZN	Wolverton	Railcare

DEPOT TYPE ABBREVIATIONS

CARMD	Carriage Maintenance Depot
T&RSMD	Traction and Rolling Stock Maintenance Depot.
TMD	Traction Maintenance Depot.
TMD (D)	Traction Maintenance Depot (Diesel).
SD	Servicing Depot.